Springer Theses

Recognizing Outstanding Ph.D. Research

For further volumes:
http://www.springer.com/series/8790

Aims and Scope

The series “Springer Theses” brings together a selection of the very best Ph.D. theses from around the world and across the physical sciences. Nominated and endorsed by two recognized specialists, each published volume has been selected for its scientific excellence and the high impact of its contents for the pertinent field of research. For greater accessibility to non-specialists, the published versions include an extended introduction, as well as a foreword by the student’s supervisor explaining the special relevance of the work for the field. As a whole, the series will provide a valuable resource both for newcomers to the research fields described, and for other scientists seeking detailed background information on special questions. Finally, it provides an accredited documentation of the valuable contributions made by today’s younger generation of scientists.

Theses are accepted into the series by invited nomination only and must fulfill all of the following criteria

- They must be written in good English.
- The topic should fall within the confines of Chemistry, Physics, Earth Sciences, Engineering and related interdisciplinary fields such as Materials, Nanoscience, Chemical Engineering, Complex Systems and Biophysics.
- The work reported in the thesis must represent a significant scientific advance.
- If the thesis includes previously published material, permission to reproduce this must be gained from the respective copyright holder.
- They must have been examined and passed during the 12 months prior to nomination.
- Each thesis should include a foreword by the supervisor outlining the significance of its content.
- The theses should have a clearly defined structure including an introduction accessible to scientists not expert in that particular field.

Susanne Mertens

Background Processes in the Electrostatic Spectrometers of the KATRIN Experiment

Doctoral Thesis accepted by
the Karlsruhe Institute of Technology, Germany

Springer

Author
Dr. Susanne Mertens
Nuclear Science Division
Lawrence Berkeley National Laboratory
Berkeley, CA
USA

Supervisor
Prof. Dr. Guido Drexlin
Department of Physics
KIT Institute of Experimental Nuclear Physics
Eggenstein-Leopoldshafen
Germany

ISSN 2190-5053 ISSN 2190-5061 (electronic)
ISBN 978-3-319-37664-6 ISBN 978-3-319-01177-6 (eBook)
DOI 10.1007/978-3-319-01177-6
Springer Cham Heidelberg New York Dordrecht London

Printed on acid-free paper

Springer is part of Springer Science+Business Media (www.springer.com)

Cosmic Gall

Neutrinos, they are very small.
They have no charge and have no mass
And do not interact at all.
The earth is just a silly ball
To them, through which they simply pass,
Like dustmaids down a drafty hall
Or photons through a sheet of glass.
They snub the most exquisite gas,
Ignore the most substantial wall,
Cold-shoulder steel and sounding brass,
Insult the stallion in his stall,
And, scorning barriers of class,
Infiltrate you and me! Like all
And painless guillotines, they fall
Down through our heads into the grass.
At night, they enter at Nepal
And pierce the lover and his lass
From underneath the bed – you call
It wonderful; I call it crass.

John Updike (1963)

Supervisor's Foreword

Neutrinos continue to be the most mysterious, and arguably the most fascinating particles of the Standard Model as their intrinsic properties such as absolute mass scale and CP properties are unknown. Ever since the first detection of neutrino oscillations more than a decade ago, the fact that neutrinos have a small but non-zero mass points to new physics beyond the Standard Model, as well as to a distinct role of primordial neutrinos from the Big Bang in the evolution of large-scale structures in the universe. The open question of the absolute neutrino mass scale will be addressed with unprecedented accuracy by the Karlsruhe Tritium Neutrino (KATRIN) experiment, currently under construction at Tritium Laboratory Karlsruhe at the KIT Campus north site. The challenging task to push the experimental sensitivity down to a neutrino mass of 200 meV requires the development of an almost background-free, high-precision spectroscopy of 18.6 keV electrons from -decay of molecular tritium. This is realised by the so-called MAC-E filter technique, where electrons from a high intensity tritium source are guided by strong magnetic fields to an electrostatic retarding spectrometer for precision energy analysis, so that the spectrum can be recorded in a detector in an integral mode.

The thesis of Susanne Mertens has been focused on the spectrometer part of KATRIN and background processes therein. In the course of this thesis, various background sources such as large and small Penning traps, as well as nuclear decays from single radon atoms could be fully characterized for the first time. This has been achieved by a detailed measurement program with the KATRIN pre-spectrometer, and by comparing experimental results with KASSIOPEIA, a newly developed electromagnetic field calculation and particle tracking code of unprecedented precision. Most importantly, however, the background in the spectrometer could be reduced by more than five orders of magnitude by eliminating Penning traps and by developing a completely new background reduction method by stochastically heating trapped electrons by electron cyclotron resonance (ECR).

It is very rare indeed that within a single thesis substantial contributions to three major breakthroughs can be pooled: the development of a new Monte Carlo program for field calculations and particle tracking, the successful reduction of background by five orders of magnitude down to a negligible level by novel methods, and the extrapolation of these breakthrough results to further refine the sensitivity of a next-generation direct neutrino mass experiment such as KATRIN.

Neutrino physics is one of the most challenging experimental disciplines not only in particle physics, but also prone to a multitude of background issues. This thesis has demonstrated beautifully that the obstacles and challenges in measuring the absolute mass scale of neutrinos can be met successfully, if novel experimental tools (ECR) and novel computing methods (KASSIOPEIA) are combined to enable tritium β-spectroscopy almost free of background.

Eggenstein-Leopoldshafen, March 2013 Prof. Dr. Guido Drexlin

Abstract

The knowledge of the absolute neutrino mass scale is an extremely important input parameter both for cosmological models and for the fundamental understanding of the nature of particle masses. So far, however, only a lower limit of 0.05 eV, provided by neutrino oscillation experiments, and an upper bound of 2 eV from direct neutrino mass experiments is limiting the allowed parameter range. The Karlsruhe Tritium Neutrino Experiment KATRIN is designed to determine the effective electron anti neutrino mass with a sensitivity of 200 meV (90 % C.L.). This is achieved by a model independent technique based on the kinematics of tritium decay. An essential prerequisite to achieve the design sensitivity is an ultra-low background level.

The focus of this work lies on the investigation of the main background sources of the KATRIN spectrometers. With the Monte Carlo simulation software KASSIOPEIA, developed in the frame of this thesis, the three main background production mechanisms were investigated: Muon-induced background, background due to penning traps, and background arising from stored electrons. KASSIOPEIA was used to model different background production mechanisms, to predict background rates and characteristics at the main spectrometer, and finally to analyze their impact on the neutrino mass sensitivity of KATRIN.

The investigations revealed that the dominant background source expected at KATRIN are stored electrons arising from ^{220}Rn and ^{219}Rn α-decay and Tritium β-decay in the volume of the main spectrometer. This is highlighted most notably by the fact that a single nuclear decay can produce an enhanced background level for up to 10 h.

To eliminate the background arising from stored electrons, a novel method based on stochastic heating by using the technique of Electron Cyclotron Resonance (ECR) was tested successfully at the pre-spectrometer. These measurements as well as corresponding simulations demonstrate explicitly that a high frequency field tuned to the cyclotron frequency of the stored electrons breaks their storage condition by stochastic heating within a time period of less than 10 ms. This method, if implemented successfully at the up-coming measurement at the main spectrometer, will enable a neutrino mass measurement with an almost background free spectrometer.

Acknowledgments

I would like to thank all the people who supported me during my Ph.D. thesis. In particular, I would like to thank:

- Prof. Dr. G. Drexlin for giving me the opportunity to work at KATRIN and for his invaluable supervision and support,
- Prof. Dr. M. Feindt for accepting to be the second reviewer of this thesis,
- Dr. Ferenc Glück for his great help and advice throughout my thesis,
- Benjamin Leiber for his patient help with computer problems and for his support with the work on our papers,
- Nancy Wandkowsky for her great support with all the numerous and time-consuming Monte Carlo simulations,
- Daniel Furse for his exceptional help with physics problems and all his great effort to make Kassiopeia a fantastic code,
- Dr. Joachim Wolf for his support, especially with vacuum calculations,
- Stefan Görhardt, Dr. Sascha Wüstling, Armen Beglarian, Hans Skacel, Dr. Klaus Schlösser, and many others for making the ECR measurement at the pre-spectrometer possible,
- Joe Formaggio, Noah Oblath, John Barrett, Daniel Furse, and T. J. Corona for the great time spent at MIT,
- Nancy Wandkowsky, Ferenc Glück, Daniel Hilk and Stefan Groh, for proof-reading my thesis,
- All colleagues from the Institut für Kernphysik for the nice working atmosphere, which I will miss,
- Antoine Tollenaere for his support all along my work, and his LaTeX and Emacs skills that saved me a lot of time,

and my friends and family for always being there for me.

Contents

Chapter 1
Preamble

Previous single β-decay experiments have revealed that the neutrino is at least five orders of magnitude lighter than its charged leptonic partner, the electron. On the other hand, neutrino oscillations have proven that the neutrino is not massless, as assumed in the Standard Model. Owing to their interference nature, oscillation experiments are only sensitive to mass differences between the neutrino mass eigenstates, but not the absolute neutrino mass scale. Assuming the lightest mass eigenstate to be of zero mass, a lower mass limit for the heaviest mass eigenstate of about 0.05 eV can be given.

Since the neutrino might be a Majorana particle (i.e. "its own antiparticle") its mass generation mechanism could be different from that of all other fermions of the Standard Model. The exact knowledge of the absolute neutrino mass scale will be of prime importance for finding out which neutrino mass model is realized in nature. This feature of the microcosm of particle physics is also of major interest in cosmology, as the universe is filled with 336 relic neutrino per cm^3 from the Big Bang. Even if neutrinos have a rather small mass their impact on the structure formation in the early universe is large. Correspondingly, the neutrino mass is a very important input parameter for cosmological model calculations.

The **K**arlsruhe **T**ritium **N**eutrino (KATRIN) experiment is a next generation, large-scale, single β-decay experiment. It is designed to directly measure the effective electron antineutrino mass with a sensitivity of 200 meV. The experiment will analyze the shape of the high energy end of the tritium β-spectrum. A non-zero neutrino mass reduces the endpoint energy and distorts the spectrum, especially in the vicinity of the endpoint. This spectrum will be analyzed with a 24×10 m electrostatic spectrometer combined with Magnetic Adiabatic Collimation (MAC-E-Filter). To reach the design sensitivity of $m_{\nu_e} < 200$ meV (90 % C.L.), high energy resolution, high signal count rates and ultra-low background have to be met.

One of the main challenges of KATRIN is that only a tiny fraction of 10^{-13} of all tritium β-decay electrons are created in the energy region of interest (i.e. 1 eV below the endpoint). Even with an unsurpassed rate of 10^{11} tritium decays per second the resulting signal rate at the endpoint is less than 0.01 counts/s. Simulations show that

S. Mertens, *Background Processes in the Electrostatic Spectrometers of the KATRIN Experiment*, Springer Theses, DOI: 10.1007/978-3-319-01177-6_1,

the KATRIN design sensitivity of 200 meV can only be reached with a background level which is of the same order of magnitude or smaller.

The objectives of this work are the investigation of the main background processes occurring in the KATRIN spectrometers and the testing of a novel active background reduction mechanism based on Electron Cyclotron Resonance (ECR).

An essential prerequisite for background modeling and suppression was the development of a full Monte Carlo simulation software for KATRIN in collaboration with the simulation task group. In this context, the main focus of this work was on the development of the general structure of the code and the implementation of a ultra-precise tracking module of charged particles in electromagnetic fields.

As KATRIN is not an experiment located in a deep underground laboratory it is exposed to the undiminished cosmic muon flux from the atmosphere. The most important countermeasure against cosmics is provided by the magnetic shielding, which suppresses electrons created by cosmic muons hitting the large spectrometer. One of the goals of this work was to investigate the impact of non-axially symmetric magnetic field components which disturb this magnetic shielding and to estimate the resulting muon induced background rate. To study the influence of non-perfect shielding on muon induced electrons, Monte Carlo simulations, incorporating a realistic model of all magnetic field sources present at KATRIN, needed to be implemented.

Apart from muon induced reactions intrinsic background sources from Penning traps can contribute significantly to the overall background rate. Measurements at the pre-spectrometer have revealed that small Penning traps can cause background rates of several kHz. One of the aims of this work was to locate and characterize the responsible Penning trap by precision electromagnetic field calculations and to develop a phenomenological model to explain how a this Penning trap of a cross sectional area of only 0.75 cm^2 can cause such high background rates.

The main focus of this thesis was on the study of background due to stored electrons following radon and tritium decays in the volume of the electrostatic spectrometers. The challenge here was to explain the measured rates and topologies at the pre-spectrometer in the framework of a detailed background model. Based on excellent agreement between experiment and Monte Carlo, the expected background characteristics at the main spectrometer and the impact on the neutrino mass sensitivity were investigated. It could be shown that the vacuum design of KATRIN will have to be drastically modified in order not to exceed the design background.

However, even if an improved vacuum system is implemented, active background reduction methods are needed. The goal of this work was to experimentally test an active background reduction technique based on stochastic heating of stored electrons by ECR. By combining measurements with accompanying simulations, detailed predictions of the efficiency of the method at the main spectrometer were carried out to investigate the possibility of background-free neutrino mass measurement.

Chapter 2
Introduction to Neutrino Physics

In neutrino physics many exciting discoveries have been made in the last decades. Most importantly, the discovery of neutrino oscillations, first indicated by the Homestake experiment [1, 2] and finally validated by Super-Kamiokande in 1998 [3] and SNO in 2001 [4, 5]. With the help of an impressive suite of atmospheric, solar, reactor and accelerator neutrino experiments most of the free parameter space of neutrino oscillation could be determined. However, only in June 2011 a first hint of a non-zero value of the remaining undetermined mixing angle θ_{13} was found by T2K in Japan [6] and more recently by the Double Chooz experiment [7].

Despite many fundamental discoveries in neutrino physics the elusive "ghost particle of the universe" still provides many burning open questions. In the Standard Model the neutrino is assumed to be massless. The discovery of neutrino oscillations, however, has proven that the neutrino is not massless in nature. In this context, single β-decay experiments were aimed to determine the absolute neutrino mass and could show that it is at least lighter by a factor of 10^5 compared to the electron mass. This factor gives rise to the puzzling question of how the neutrino acquires mass.

A closely related open question is the nature of the neutrino with respect to the CP-symmetry operator, i.e. whether it is a Majorana particle or a Dirac particle. Experiments looking for neutrinoless double β-decay are ideally suited to answer this question.

Neutrino oscillations depend on the mass splittings Δm_{ij}^2 between the three mass eigenstates of the neutrino. The parameters Δm_{12}^2 and Δm_{23}^2 have been measured by several oscillation experiments to rather good accuracy. However, so far, the sign of Δm_{23}^2 is still undetermined. Therefore, the hierarchy of the neutrino masses is still not known.

Finally, the absolute mass scale of the neutrino is so far only constrained by single β-decay experiments, with the most sensitive experiments at Mainz and Troitsk setting an upper limit of 2 eV [8–11]. Cosmology and the search for neutrinoless double β-decay can provide more stringent bounds, however, these results are much more model dependent. As the knowledge of the absolute neutrino mass scale is

S. Mertens, *Background Processes in the Electrostatic Spectrometers of the KATRIN Experiment*, Springer Theses, DOI: 10.1007/978-3-319-01177-6_2,

an extremely important input both for cosmological models and for the fundamental understanding of the nature of the mass, all three methods complement each other.

This chapter gives an overview of neutrino physics. In Sect. 2.1 a brief history of the discovery of the neutrinos is given. In Sect. 2.2 the principle and status of neutrino oscillations is presented. Section 2.3 describes the minimal extensions of the Standard Model, that allow for introducing a neutrino mass term. Finally, the three major approaches to determine the neutrino mass will discussed in Sect. 2.4.

2.1 The Discovery of the Neutrino

The neutrino was part of theoretical physics long before it was actually detected. After Pauli postulated it, a period of 26 years passed until it was detected by project Poltergeist, fortunately still during Pauli's lifetime. In the following these two landmarks, Pauli's postulate and the experimental discovery of neutrinos of all flavors, will be outlined.

2.1.1 Postulation of the Neutrino by Pauli

When the energy spectrum of β-decay electrons was first investigated in 1914 by James Chadwick [12], he expected to measure a mono-energetic line of the electron, owing to the then assumed two-body decay nature. However, a continuous spectrum was observed, as shown in Fig. 2.1. This observation is in contradiction to energy and angular momentum conservation if a two-body decay is assumed. In his famous Letter dated to the 4th of December 1930, Pauli [13] postulated the existence of a neutral spin $\frac{1}{2}$ particle, the "neutron", which is produced in β-decays alongside the charged electron. In that case the β-decay can be described as three-body decay, and the contradictions to conservation laws are resolved. When the neutron was two years

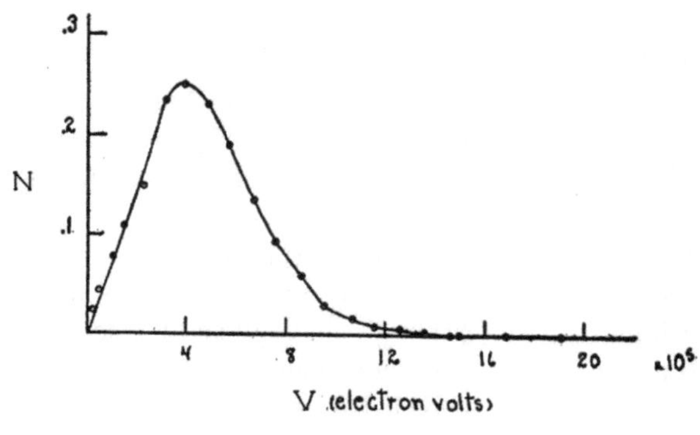

Fig. 2.1 Continuous energy spectrum of the beta electrons from radium decay [15]

later found, however, it was clear that due to its large mass this particle could not be the missing neutral particle in the β-decay. In 1934, E. Fermi, who theoretically described the β-decay, introduced the name "neutrino" for Pauli's hypothetical particle [14]. Fermi derived an expression for the shape of the electron energy spectrum, and, from comparison with available data deduced that the neutrino mass must be either zero or much smaller than the electron mass.

2.1.2 First Detection of a Neutrino

The first detector that "saw" the neutrino was the detector "Herr Auge" (see Fig. 2.2a) located at the Hanford reactor site as the centerpiece of the famous project Poltergeist by C. Cowan and F. Reines. However, the background in this 1954 pioneering experiment was still overwhelming the signal. Shortly after these initial studies, C. Cowan and F. Reines could definitively prove the existence of the neutrino with an improved detector (see Fig. 2.2b) at the Savannah river reactor [16, 17]. The neutrino was detected by the classical inverse β-decay

$$\overline{\nu}_e + p \rightarrow n + e^+. \qquad (2.1)$$

The Savannah River detector consisted of liquid scintillator tanks inter spaced with a Cadmium-loaded (Cd) water target. The positron from Eq. (2.1) annihilates with an electron, giving rise to a prompt signal of two back to back gammas. The neutron is thermalized on a timescale of milliseconds and finally captured by Cd, thereby releasing gammas when the excited Cd-state decays to the ground state

(a) (b)

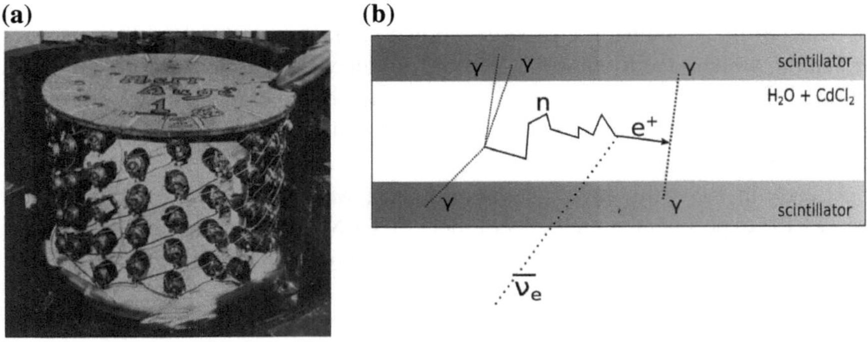

Fig. 2.2 First neutrino detector and detection principle. **a** Photograph of the first neutrino detector named "Herr Auge", *Source* [16]. **b** Detection principle: the antineutrino $\overline{\nu}_e$ from the reactor hits a free proton in the $H_2O + CdCl_2$ target thereby producing a positron and a neutron. The positron annihilates, while the neutron is captured on cadmium after moderation. The subsequent gammas are detected by Compton scattering in the liquid scintillation detector. The scintillation light is detected by photomultipliers

(see Fig. 2.2b). These gammas are detected by their subsequent Compton scattering. The scintillation light is detected by photomultiplier tubes surrounding the detector volume. The promt and the delayed light signal represent a distinct coincidence signature of a neutrino interacting in the detector. Long after this crucial discovery, the Nobel Prize was given to F. Reines in 1995 for the first detection of the neutrino.

2.1.3 Discovery of ν_μ and ν_τ

The second neutral lepton flavor state, the muon neutrino ν_μ was found by L.M. Lederman, M. Schwartz and J. Steinberger in 1962 at the Brookhaven Alternating Gradient Synchrotron (AGS) [18]. There, neutrinos from pion decay

$$\pi^+ \rightarrow \mu^+ + \nu_\mu \tag{2.2}$$

were investigated. To identify the ν-flavor associated with the π-decay, the flavor of the charged lepton, being produced via a charged current interaction in the detector, had to be detected. For this purpose a spark chamber made of 10 t of aluminum was used. As only tracks from muons and no electronic showers were observed, the conclusion was, that the neutrinos produced together with a muon, i.e. muon neutrinos, are intrinsically different from electron (anti-)neutrinos. For the discovery of a second kind of neutrino the three were awarded with the Nobel prize in 1988.

Finally, in the year 2000, the tau neutrino ν_τ was detected in the Direct Observation of Nu Tau (DONUT) experiment at Fermilab [19]. In this experiment a beam of 800 GeV protons impinged on to a tungsten target to create a particle shower. A small fraction of the particles decays into τ leptons, which in turn produce ν_τ when they decay. With a massive shielding all particles but the ν_τ were removed. The ν_τ's are detected in an emulsion lead target. When interacting in the lead they produce a τ lepton, which, due to its short lifetime, produces a short track in the emulsion. All decay products, but the neutrino, leave a track in the emulsion with a different direction than the τ. The famous signature of τ neutrinos is therefore a "kink" in the particle tracks.

Already in 1989, the three generation picture with $N_\nu = 3$ was established by the ALEPH experiment at the LEP collider at CERN. This was based on the precise measurement of the invisible width of the Z boson resonance at 91 GeV [20, 21].

2.2 Neutrino Oscillations

The discovery of neutrino oscillations is a crucial milestone for neutrino physics with far reaching implications for particle physics and cosmology. It proves that the neutrinos are not massless, since it requires the neutrino mass eigenstates to have different masses. Assuming the lightest neutrino ν_1 being (almost) massless, the measurement of the mass splittings gives a lower mass bound. In this section the discovery of neutrino oscillation, the standard theoretical treatment and the neutrino oscillations in matter will be described.

2.2.1 First Discovery

In the second half of the 1960s physicists and nuclear chemists started seriously thinking about measuring neutrinos from the sun. In nuclear fusion processes, i.e. in the pp-chain and the sub-dominant CNO-cycle [22], neutrinos with exclusively electron-type flavor are produced. The flux of solar neutrinos at the distance of the earth equals to about 60 billion neutrinos per cm^2 and second. Figure 2.3 shows the energy spectrum of the neutrinos being produced in the solar core. In contrast to reactors, the pp and CNO cycles in the sun only produce electron neutrinos ν_e (i.e. no electron anti neutrinos $\overline{\nu}_e$).

In the first solar neutrino experiment, the Homestake experiment led by Ray Davis Jr. [1, 2], the radiochemical detection technique was pioneered.Homestake was based on a tank of 600 t of perchloroethylene, containing the isotope ^{37}Cl as neutrino target. The ^{37}Cl nucleus, when interacting with a ν_e from the sun

$$\nu_e + {}^{37}\text{Cl} \leftrightarrow e^- + {}^{37}\text{Ar}, \tag{2.3}$$

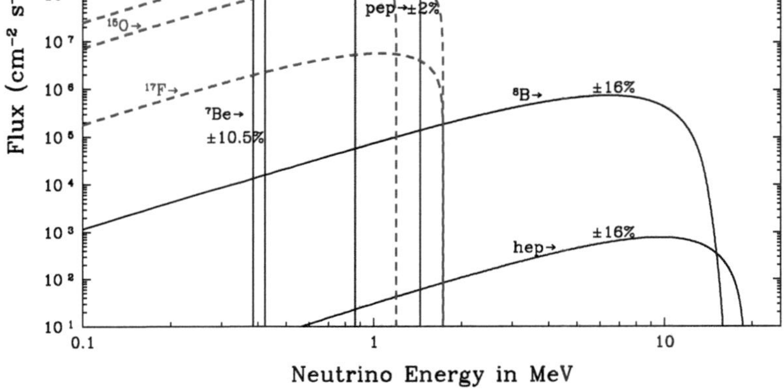

Fig. 2.3 Solar neutrino spectrum. In this plot the flux of neutrinos from the pp-chain (*solid line*) and the CNO cycle (*dashed lines*) are shown as a function of energy. The integral spectrum is measured down to 0.2 MeV by e.g. Gallex and SAGE. Experiments measuring a differential spectrum have a higher threshold. SNO detects $E_\nu > 6$ MeV, Borexino [23, 24] can reach below 1 MeV [25]

is transmuted into an excited ^{37}Ar atom. After a measuring interval of several weeks, the produced ^{37}Ar atoms are separated from the target material. Their subsequent decay via electron capture to excited levels of ^{37}Cl which de-excites via Auger emission is counted in a proportional counter. Surprisingly, Davis et al. found fewer neutrino interactions than predicted by the Standard Solar Model. The deficit was confirmed by other experiments, e.g. Gallex/GNO, SAGE and Kamiokande. These observations established the solar neutrino problem. The deficit was either pointing to an incorrect solar model or the fact that the neutrinos undergo a change of flavor on their way from the center of the sun to the detection on earth. The experiments mentioned above were almost exclusively sensitive to the electron flavor type only and therefore were not able to detect neutrinos of other flavors, which are expected if neutrinos are subject to neutrino oscillations.

A similar observation, yet in the GeV energy range, was made by the Super-Kamiokande experiment [3] which measured a deficit of ν_μ from the atmosphere. In the atmosphere neutrinos are mainly produced via pion decay and subsequent muon decay, leading to a 2:1 ratio of ν_μ to ν_e. Interestingly, the deficit of ν_μ was maximal for those neutrinos passing through the earth, i.e. at a maximal distance to the source.

To definitively test the neutrino oscillation hypothesis in the solar sector the Sudbury Neutrino Observatory (SNO) was built [4, 5]. SNO was designed to provide sensitivity to all neutrino flavors. In the first phase, 1,000 t of pure heavy water were used as detector medium. The D_2O target allows for elastic scattering (ES) off electrons, as well as, neutral (NC) and charged current (CC) interactions on deuterium:

$$\nu_x + e \rightarrow \nu_x + e^- \quad \text{(CC for } \nu_e \text{ only, NC for all flavors)} \tag{2.4}$$

$$\nu_e + d \rightarrow p + p + e^- \quad \text{(CC for } \nu_e \text{ only)} \tag{2.5}$$

$$\nu_x + d \rightarrow p + n + \nu_x \quad \text{(NC for all flavors)} \tag{2.6}$$

Each of the different channels gives rise to a specific signal. Therefore it is possible to measure the total neutrino flux and the electron neutrino flux separately. To validate and refine the result the fluxes were cross-checked with different detection schemes for neutrons from the NC reactions, based on pure D_2O (Phase 1), salt water (Phase 2) and ^{3}He (Phase 3) as medium for neutron capture. It could indeed be shown that the total neutrino flux from the sun is conserved, leading to the conclusion that the neutrino flavor changes on the way from the sun to the earth. Figure 2.4 shows the flux results for ν_μ, ν_τ and ν_e.

2.2.2 Theoretical Description

The phenomenon of neutrino oscillations arises from the fact that neutrino mass eigenstates are not identical to the flavor eigenstates, i.e. the neutrino state coupling to weak interaction is not equal to the state propagating. This mixing can be considered analogous to the well-known mixing in the quark sector, where quarks however

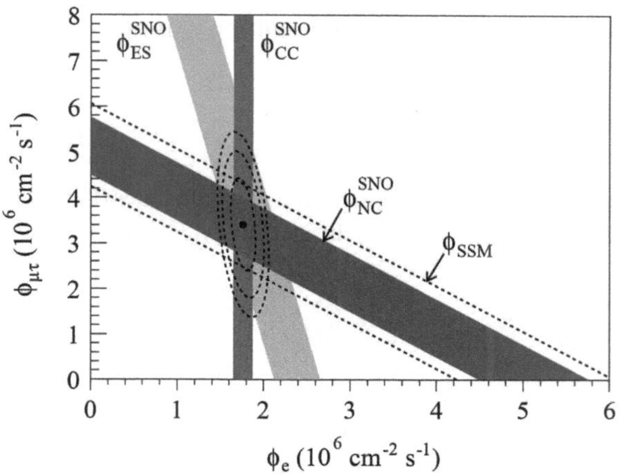

Fig. 2.4 Flavor composition of ^{8}**B** neutrinos from the sun. The *three solid bands* show the fluxes measured via different reactions in SNO. The charged current reaction (CC) determines the electron neutrino flux, while the neutral current reaction is not sensitive to the flavor and determines the total flux. The total flux is in agreement with the prediction (*dashed line*). The intersection of the CC, NC and ES bands indicates that the flux is composed of 1/3 ν_e and 2/3 ν_μ and ν_τ [26]

cannot be detected as freely propagating particles due to confinement. A better comparison and analogy is thus given by the K and B meson oscillation [27].

A neutrino flavor eigenstate ν_α, with $\alpha = e, \mu, \tau$ is defined as a superposition of neutrino mass eigenstates ν_i, with $i = 1, 2, 3$.

$$\begin{pmatrix} \nu_e \\ \nu_\mu \\ \nu_\tau \end{pmatrix} = \begin{pmatrix} U^*_{e1} & U^*_{e2} & U^*_{e3} \\ U^*_{\mu 1} & U^*_{\mu 2} & U^*_{\mu 3} \\ U^*_{\tau 1} & U^*_{\tau 2} & U^*_{\tau 3} \end{pmatrix} \begin{pmatrix} \nu_1 \\ \nu_2 \\ \nu_3 \end{pmatrix} \tag{2.7}$$

where U is called the Pontecorvo-Maki-Nakagawa-Sakata (PMNS) matrix. It contains three mixing angles θ_{ij} and one non-trivial complex Dirac phase (δ_D). In addition there are 2 complex Majorana phases (δ_M). The phases can cause CP violation and are of relevance in double β-decay. The PMNS matrix is most commonly factorized in the form

$$U = \begin{pmatrix} 1 & 0 & 0 \\ 0 & c_{23} & s_{23} \\ 0 & -s_{23} & c_{23} \end{pmatrix} \begin{pmatrix} c_{13} & 0 & s_{13}e^{-i\delta_D} \\ 0 & 1 & 0 \\ -s_{13}e^{-i\delta_D} & 0 & c_{13} \end{pmatrix} \begin{pmatrix} c_{12} & s_{12} & 0 \\ -s_{12} & c_{12} & 0 \\ 0 & 0 & 1 \end{pmatrix} \begin{pmatrix} e^{i\delta_{M1}} & 0 & 0 \\ 0 & e^{i\delta_{M2}} & 0 \\ 0 & 0 & 1 \end{pmatrix}, \tag{2.8}$$

where $s_{ij} = \sin(\Theta_{ij})$ and $c_{ij} = \cos(\Theta_{ij})$.

Considering a case in which an electron neutrino ν_e is produced, the state at $t = 0$ can be written as

$$|\nu\,(t > 0)\rangle = |\nu_e\rangle = U^*_{e1}\,|\nu_1\rangle + U^*_{e2}\,|\nu_2\rangle + U^*_{e3}\,|\nu_3\rangle\,. \tag{2.9}$$

The mass eigenstates ν_i are the physical states (eigenstates of the free Hamiltonian) that propagate through space with a definite energy E_i and momentum p_i. After a certain time $t > 0$ the state evolves to

$$|\nu\,(t > 0)\rangle = U^*_{e1}e^{-iE_1t}\,|\nu_1\rangle + e^{-iE_2t}U^*_{e2}\,|\nu_2\rangle + e^{-iE_3t}U^*_{e3}\,|\nu_3\rangle \neq |\nu_e\rangle\,. \tag{2.10}$$

This superposition of mass eigenstates is not necessarily a flavor eigenstate. Therefore one finds a non-vanishing probability to measure the neutrino in a different flavor than at origin $t = 0$. The neutrino is always detected in a flavor eigenstate, as it interacts only by the weak force.

Using the fact that each mass eigenstate can itself be expressed as a superposition of flavor eigenstates one can more generally express a flavor state ν_α at a time t as

$$|\nu_\alpha\,(t)\rangle = \sum_k U^*_{\alpha k}e^{-iE_kt}\,|\nu_k\rangle \tag{2.11}$$

$$= \sum_{\beta=e,\mu,\tau} \underbrace{\left(\sum_k U^*_{\alpha k}e^{-iE_kt}U^*_{\beta k}\right)}_{A_{\nu\alpha\to\nu_\beta(t)}} |\nu\beta\rangle, \tag{2.12}$$

where

$$|\nu_k\rangle = \sum_{\beta=e,\mu,\tau} U_{\beta k}\,|\nu_\beta\rangle \tag{2.13}$$

was used. The probability P to find a certain flavor state ν_β is given by the projection of the state $|\nu_\alpha\,(t)\rangle$ onto the flavor eigenstate $|\nu_\beta\rangle$.

$$P\left(\nu_{\alpha\to\beta}\,(t)\right) = \left|\langle\nu_\beta\,|\nu_\alpha\,(t)\rangle\,\right|^2 = \left|A_{\nu_\alpha\to\nu_\beta}\,(t)\right|^2 = \left|\sum_k U^*_{\alpha k}e^{-iE_kt}U_{\beta k}\right|^2 \tag{2.14}$$

$$= \sum_{k,j} U^*_{\alpha k}U_{\beta k}U_{\alpha j}U^*_{\beta j}e^{-i(E_k-E_j)^t} \tag{2.15}$$

Using an ultra relativistic approximation with the assumption $p_k = p = E$ one derives the following standard expression for P:

$$P(\nu_{\alpha\to\beta}(L/E)) = \sum_{k,j} U^*_{\alpha k}U_{\beta k}U_{\alpha j}U^*_{\beta j}e^{-i\frac{\Delta m^2_{jk}L}{2E}} \tag{2.16}$$

with $\Delta m_{ik}^2 = m_i^2 - m_k^2$ corresponding to the mass splittings, L denoting the distance between source and detector, and E corresponding to the energy of the neutrino. One can see that the oscillation probability P is determined by the mixing parameters of the PMNS matrix and the two mass splittings.

Since one of the three mixing angles θ_{13} is found to be small the generic 3ν mixing case is reduced to decoupled two neutrino flavor oscillation. In that case the transition probability is simplified to the generic 2ν case:

$$P_{\nu_e \rightleftarrows \nu_\mu}(L/E) = \sin^2(2\Theta)\sin^2\left(\frac{\Delta m^2 L}{2E}\right) \tag{2.17}$$

This equation nicely shows that the amplitude of the oscillation probability $\sin^2(2\Theta)$ depends on the corresponding mixing angle Θ, while the mass spitting Δm^2 determines the frequency. A more general description of neutrino oscillations, which makes no use of the ad hoc assumptions of the standard approach, is based on a quantum field theoretical wave packet model [28, 29].

2.2.3 Determination of Oscillation Parameters

To determine the full set of ν-mixing parameters different neutrino sources have to be used. These include natural sources like the atmosphere and the sun, and man made sources, like reactors and accelerators. The specific L/E ratios are tuned to the mass splitting Δm^2 and mixing angle Θ being investigated.

Solar neutrino experiments are most sensitive to the mixing angle Θ_{12}, also called the solar mixing angle, whereas atmospheric oscillation experiments are most sensitive to the mixing angle Θ_{23} (atmospheric mixing angle). More recently, the main emphasis of oscillation experiments has shifted to tune the baseline of reactor or accelerator experiments to optimize the sensitivity to the unknown parameter θ_{13}.

By adjusting the L/E ratio, an experiment can measure the oscillatory nature of the flavor transitions. A prime example, shown in Fig. 2.5, is given by the measurement results of the Kamioka Liquid Scintillator Antineutrino Detector (KamLAND) [30]. KamLAND is a long-baseline experiment at the Kamioka Laboratory in Japan. It observes reactor neutrinos at an average distance of 180 km (mean of all Japanese reactors) and is therefore measuring the "solar" mixing angle Θ_{12}. The result nicely shows the oscillation pattern given in Eq. (2.17).

The best fit result of all current experiment is [31]:

$$\sin^2(2\Theta_{12}) = 0.87 \pm 0.03 \tag{2.18}$$

$$\Delta m_{12}^2 = (7.59 \pm 0.20) \cdot 10^{-5}\ \mathrm{eV}^2 \tag{2.19}$$

$$\sin^2(2\Theta_{23}) > 0.92 \tag{2.20}$$

$$|\Delta m_{23}^2| = (2.43 \pm 0.13) \cdot 10^{-3}\ \mathrm{eV}^2 \tag{2.21}$$

$$\sin^2(2\Theta_{13}) < 0.15 \quad (90\,\%\,\mathrm{CL}) \tag{2.22}$$

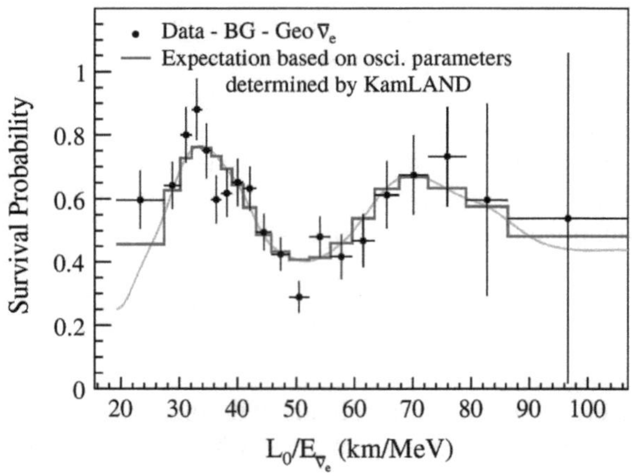

Fig. 2.5 Result of KamLAND. Survival probability of reactor neutrinos $\overline{\nu}_e$ as a function of distance over energy [30]. By considering different energy regimes of the neutrinos the L/E ratio is varied and an oscillation pattern becomes visible

The "atmospheric" mixing angle Θ_{23} is almost maximal (i.e. 45°), Θ_{12} is very large and Θ_{13} is very small. Recent results from T2K reveal a parameter space $0.03\ (0.04) < \sin^2(2\Theta_{13}) < 0.28\ (0.34)$ for no CP violation and normal (inverted) hierarchy at 90 % CL [6]. This has been corroborated by very recent results from Double Chooz [7].

For Δm^2_{23} only the absolute value is known, hence the generic mass hierarchy of the neutrinos is not known. Oscillations in matter are sensitive to the sign of the mass splitting, therefore the sign of Δm^2_{12}, as deduced from solar neutrinos travelling in solar matter, is known.

2.2.4 Oscillation in Matter

The oscillation pattern is changed in the presence of matter, which is described by the so-called Mikheyev-Smirnov-Wolfenstein (MSW) effect [32]. The matter of the sun consists of baryonic matter (i.e. quarks) and electrons. Elastic scattering of neutrinos off nuclei is the same for all neutrino flavors. In the elastic scattering off electrons the neutral current component is identical for neutrinos of all flavors, however, the charged current component enters only for the electron neutrino. This flavor distinction results in an effective electron neutrinos mass and hence in a modification of the oscillation pattern.

The ν_e's "feel" an additional potential V that changes their energy E

$$E^2 \rightarrow (E + V)^2 \approx E^2 + 2EV = E^2 + 2E\sqrt{2}G_F N_e, \tag{2.23}$$

The potential V is proportional to the Fermi constant G_F and the electron density N_e. Since the Hamiltonian is changed correspondingly, also the time evolution of a flavor state is modified. In case of solar neutrino oscillations it is sufficient to consider only two flavor oscillation between ν_e and ν_x, which is a superposition of ν_μ and ν_τ. Since the neutrinos are ultra-relativistic $x = ct$, i.e. the spatial evolution is equivalent to the time evolution, the evolution of the flavor eigenstates in vacuum is given by

$$\frac{\partial}{\partial x}\begin{pmatrix}\nu_e\\ \nu_x\end{pmatrix} = H_0\begin{pmatrix}\nu_e\\ \nu_x\end{pmatrix} \tag{2.24}$$

$$= \frac{\Delta m_{12}^2}{2E}\begin{pmatrix}-\cos(2\Theta) & \sin(2\Theta)\\ \sin(2\Theta) & \cos(2\Theta)\end{pmatrix}\begin{pmatrix}\nu_e\\ \nu_x\end{pmatrix}. \tag{2.25}$$

Taking into account the flavor diagonal matter effect the evolution equation is changed to

$$\frac{\partial}{\partial x}\begin{pmatrix}\nu_e\\ \nu_x\end{pmatrix} = \left(\frac{\Delta m_{12}^2}{2E}\begin{pmatrix}-\cos(2\Theta) & \sin(2\Theta)\\ \sin(2\Theta) & \cos(2\Theta)\end{pmatrix} + V(x)\begin{pmatrix}1 & 0\\ 0 & 0\end{pmatrix}\right)\begin{pmatrix}\nu_e\\ \nu_x\end{pmatrix}. \tag{2.26}$$

Diagonalizing the Hamiltonian gives two mass eigenstates in matter which are different from the mass eigenstates in vacuum. When solving Eq. (2.26) different cases can be distinguished:

In case the neutrino is produced at very high electron densities and its energy is large, the parameter $\frac{\Delta m_{12}^2}{2E}$ can be neglected compared to V. The Hamiltonian takes a diagonal form and the neutrino at origin is mostly made of the heavy mass eigenstate in matter, as visualized in Fig. 2.6. Assuming a slow change of the density in the sun, the neutrino stays in the same mass eigenstate all the time (adiabaticity). As the density decreases with the radius, the neutrinos pass through a region where the matter effect and vacuum oscillation are equal, i.e. the diagonal terms in the Hamiltonian cancel and consequently the neutrino mass eigenstate is an equal admixture of both flavors (level crossing). As the density drops to zero at the solar surface the mass eigenstate in matter eventually turns into a mass eigenstate in vacuum. No further oscillations occur in vacuum since the state is a pure mass eigenstate. The probability to measure an electron neutrino on the earth is fixed and given by the projection of the mass eigenstate on the electron flavor, which is approximately 1/3 (see Fig. 2.7). If the neutrino has less than 1 MeV the matter effect can be neglected compared to the vacuum oscillation term. This energy dependence has been verified by a suite of experiments, most notably by Borexino.

2.3 Theory of Neutrino Mass

In the Standard Model no theoretical mechanism is foreseen to provide a mass term for neutrinos. Correspondingly, the neutrino mass is the first evidence for physics beyond the Standard Model. The fact that neutrino masses are smaller than the mass of

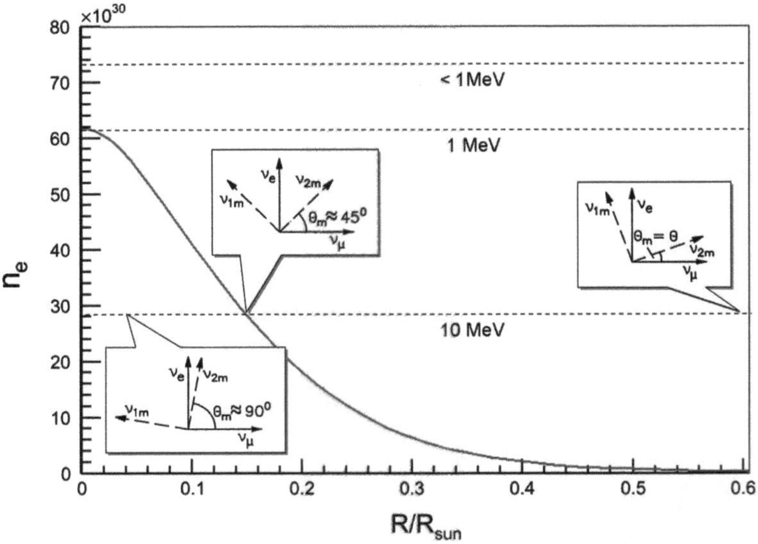

Fig. 2.6 Visualization of MSW effect. The *red line* shows electron density per m^3 as a function of solar radius. The *three horizontal lines* indicate the behavior of neutrinos of different energies in sun. For very low-energy neutrinos the matter effect can be neglected. The electron density is only relevant for neutrino energies larger than 1 MeV. For a 10 MeV ν_e, produced in the center of the sun, the matter effect V firstly dominates the vacuum oscillation term in Eq. (2.26), i.e. the Hamiltonian is approximately diagonal. The *inset on the left* visualizes how in that case the ν_e is mostly made up of the heavy mass eigenstate in matter ν_{2m}. As long as the density in the sun changes adiabatically the neutrino stays in its mass eigenstate. As the neutrino moves out of the sun, it eventually crosses an area where the matter effect and vacuum oscillation are equal, i.e. the diagonal terms in Eq. (2.26) exactly cancel. The *inset in the middle* shows that the mass eigenstate in matter is an equal admixture of both flavors. Finally, when the neutrino leaves the sun, the mass eigenstate in matter is equal to the mass eigenstate in vacuum. The angle Θ in the *inset on the right* corresponds to the vacuum mixing angle. The neutrino underwent a resonant flavor change: at production ν_{2m} is mostly ν_e, whereas in vacuum only 1/3 of $\nu_{2m} = \nu_2$ is ν_e

the charged leptons by at least 5 orders of magnitude implicates that neutrinos acquire mass by a different mechanism than the standard Higgs mechanism. In the following possible mass terms and natural mechanisms that can explain the smallness of the neutrino mass, in particular the so-called See-Saw mechanism, will be explained.

2.3.1 Possible Mass Terms in the Lagrangian

To write down a neutrino mass term analogous to the electron mass term a right chiral neutrino singlet ν_R has to be introduced to the Standard Model

$$\mathcal{L}^{\mathrm{D}}_{\mathrm{mass}} = -\sum_{l,l'} \overline{\nu}_{l'\mathrm{L}} M^{\mathrm{D}}_{ll'} \nu_{l\mathrm{R}} + \mathrm{h.c.}, \tag{2.27}$$

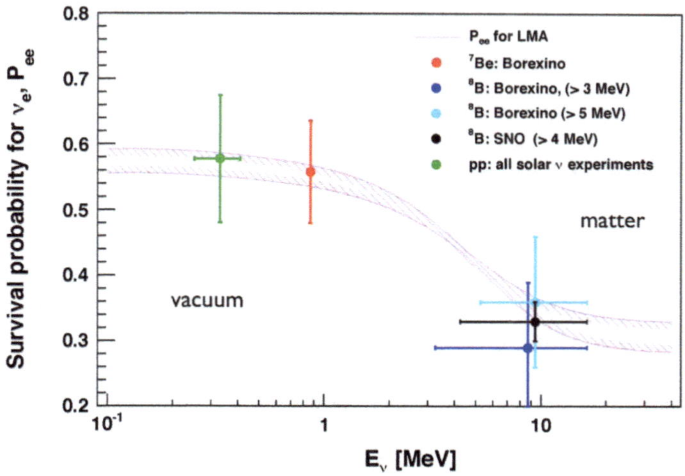

Fig. 2.7 Survival probability of electron neutrinos from the sun as a function of their energy. The figure shows that only for neutrinos of more than approximately 1 MeV the matter effect plays a role. If the vacuum oscillation dominates the survival probability is roughly 0.5

where l and l' run over all flavors and L and R denote the chirality. $\mathcal{L}^{\mathrm{D}}$, the so-called Dirac mass term, is realized via a coupling to the Higgs field. In this scenario, a neutrino mass of the right order of magnitude can only be achieved by fine tuning: The Yukawa coupling y has to be chosen unnaturally small to about $y = 10^{-13}$, which is considered not very appealing. In case the mass of the neutrino is realized in this way, the neutrino would be a so-called Dirac particle.

It is also possible to construct a mass term only from the Standard Model neutrino fields [33]

$$\mathcal{L}^{\mathrm{M,L}}_{\mathrm{mass}} = -\frac{1}{2}\sum_{l,l'} \overline{\nu}_{l'\mathrm{L}} M^{\mathrm{M,L}}_{ll'} (\nu^{C}_{l\mathrm{L}}) + \mathrm{h.c.} \tag{2.28}$$

This term violates lepton number by 2 units. If the neutrinos acquire their mass via this mechanism they are so called Majorana particles and can be considered to be their own antiparticle.

Both mass terms in Eqs. (2.27) and (2.28) must be generated with the help of a Higgsfield in order not to violate gauge invariance. However, for right chiral singlet it is possible to construct a Majorana mass term without a Higgsfield. These fields are singles under the SU(2) gauge group and therefore such a term does not violate gauge invariance.

$$\mathcal{L}^{\mathrm{M,R}}_{\mathrm{mass}} = -\frac{1}{2}\sum_{l,l'} \overline{\nu}^{C}_{l'\mathrm{R}} M^{\mathrm{M,R}}_{ll'} \nu_{l\mathrm{R}} + \mathrm{h.c.} \tag{2.29}$$

Finally, it is also possible to introduce a neutrino mass via radiative corrections or new physics, e.g. SUSY or extra dimensions [34–37].

2.3.2 See Saw Mechanism

Two ways to naturally generate a small neutrino mass can be distinguished [38, 39]. A first approach is to only consider the Majorana mass term in Eq. (2.28). To couple $(\overline{\nu}_{\mathrm{L}})^C$ and ν_{L}, which both have weak isospin $I^3_{\mathrm{W}} = +\frac{1}{2}$, a state with weak isospin $I^3_{\mathrm{W}} = -1$ has to be added, e.g. two Standard Model Higgs fields. This term then has mass dimension 5 and is interestingly the only possible dimension 5 operator made of only Standard Model particles. Since terms in the Lagrangian always have to be of dimension 4 in order to be renormalizable, such a term can only be introduced in an effective field theory approach. In that case, to reduce the dimension by 1, it has to be divided by a mass. This mass is naturally chosen to be of a high energy scale, e.g. of the GUT scale, thereby it suppresses the new physics associated with the dimension 5 term. The corresponding term in the Lagrangian

$$\mathcal{L}_{\mathrm{Majorana}} = \mathrm{g}\frac{1}{\Lambda}(\overline{\nu}_{\mathrm{L}})^C \nu_{\mathrm{L}} \phi^0 \phi^0, \tag{2.30}$$

includes a coupling constant g, a large scale Λ and two Higgs fields ϕ^0. After electroweak symmetry breaking the Higgs field acquires a non-vanishing vacuum expectation value and a Majorana mass term is generated

$$M^{\mathrm{M}} \propto \frac{1}{2}\frac{(yv)^2}{\Lambda} = \frac{m^2_{\mathrm{EW}}}{\Lambda} \tag{2.31}$$

with y denoting the Yukawa coupling and v the vacuum expectation value of the Higgs field. Introducing a neutrino mass term via a dimension 5 operator naturally provides a small neutrino mass due to the large mass scale Λ in the denominator. To generate a neutrino mass of $\propto$ 1 eV, the corresponding large mass scale Λ must be chosen to be $\propto 10^{13}$ GeV, which is interpreted to be the lepton number breaking scale. This mechanism does not naturally provide a hierarchy of the neutrino masses and therefore supports the degenerate mass regime of the neutrinos.

Another way to generate a small neutrino mass is with the help of the right chiral field. In that case the two mass terms in Eqs. (2.27) and (2.29) are considered. The Lagrangian can be written as

$$\mathcal{L}_{\mathrm{mass}} = -\frac{1}{2}\overline{\boldsymbol{\nu}}_{\mathrm{L}} M^{\mathrm{D+M,R}} (\boldsymbol{\nu}_{\mathrm{L}})^C \tag{2.32}$$

with

$$\boldsymbol{\nu}_{\mathrm{L}} = \begin{pmatrix} \nu_{\mathrm{L}} \\ (\nu_{\mathrm{R}})^C \end{pmatrix}, \tag{2.33}$$

$$M^{\mathrm{D+M,R}} = \begin{pmatrix} 0 & M^{\mathrm{D}} \\ (M^{\mathrm{D}})^T & M^{\mathrm{M,R}} \end{pmatrix} \tag{2.34}$$

When diagonalizing the mass matrix one find two mass eigenstates: a very light one m_1 and a heavy one m_2

$$m_1 \approx \frac{(M^{\mathrm{D}})^2}{M^{\mathrm{M,R}}} \tag{2.35}$$

$$m_2 \approx M^{\mathrm{M,R}} \tag{2.36}$$

As with the first mechanism (Eq. 2.31), also here small neutrino masses naturally occur. In this case also a hierarchy of neutrino masses naturally arises, due to the Dirac mass in the numerator. For a more detailed description see [40, 41].

2.4 Determination of the Neutrino Mass

A precise determination of the neutrino mass would have a large impact on both particle physics and cosmology. For particle physics the knowledge about the neutrino mass could give rise to a better understanding of the mass generating mechanism in nature. The role of the neutrino in cosmology depends on its actual mass scale, see Fig. 2.8. The mass of relic neutrinos in the universe can contribute sizably to the total matter content, and, moreover influence the structure formation in the early universe.

Three main approaches to measure the absolute neutrino mass scale are currently being explored: cosmological studies of the formation and evolution of large-scale structures (LSS), the search for neutrinoless double β-decay ($0\nu\beta\beta$), and the investigation of the kinematics of single β-decay. It should be emphasized that all methods measure a different effective neutrino mass. In the following the main ideas of the approaches will be described and the differences will be pointed out.

2.4.1 *Cosmology*

Neutrinos play an important role on cosmological scales due to their vast abundance in the universe. From the Big Bang there are about 336 relic neutrinos/cm^3 in the entire cosmos today. Consequently, even though their mass is rather small neutrinos can contribute significantly to the total matter-energy density of the universe. However, for the total dark matter in the universe to be made of neutrinos, the sum of the neutrino masses m_{ν_i} would have to be unrealistically large (~30 eV). Figure 2.8

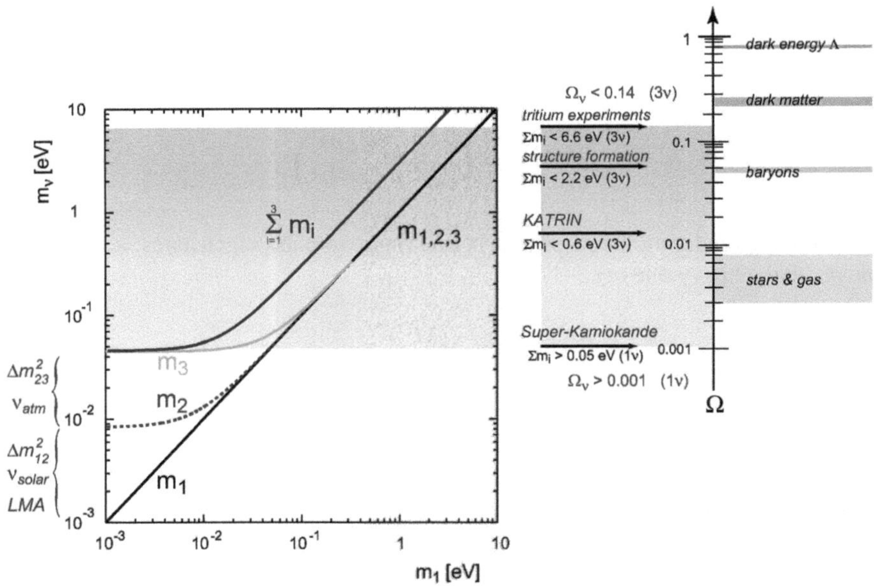

Fig. 2.8 Neutrino contribution to the total mass in the universe. The *plot on the left* shows m_{ν_i} as a function of the lightest neutrino mass eigenstate. From neutrino oscillation experiments the mass splitting are known, however, not the overall mass scale. Direct neutrino mass experiments give an upper bound on the absolute neutrino mass scale. Two extreme cases can be distinguished for the mass of the lightest neutrino mass eigenstate: it could be of the order of the mass splittings (hierarchical scenario) or of the order of the upper neutrino mass bound (degenerate scenario), where the mass splittings would play a minor role. Particle physics provides several models to generate a neutrino mass, some of which prefer the degenerate and others the hierarchical scenario. Consequently, by knowing the absolute neutrino mass scale one could distinguish between the scenarios and thereby promote some models over others. On the *right-hand side* of the figure, the possible contribution of the neutrino to the energy density of the universe is demonstrated. The *red arrows* indicate the experimental bounds on the neutrino mass [42]

shows the contribution to the total matter of the universe depending on the mass scale of the neutrino.

In the very early universe, at timescales of about 1 s ($\propto 10^9$ K), neutrinos experienced a "freeze-out" from thermal equilibrium. Thermal freeze-out of a particle occurs when the collision rate is smaller than the Hubble parameter, describing the rate of expansion of the universe. Since the freeze-out temperature of $T_\nu = 1$ MeV is very high, neutrinos at that time were ultra-relativistic. Furthermore, since the "cooling time" of the neutrinos is very slow, they influence structure formation in the form of so-called hot dark matter. Neutrinos act as "cosmic architects" by carrying energy out of clumped matter, due to their large free-streaming length. This implies that small perturbations of the matter density are washed out by neutrinos (see Fig. 2.9). Only structures of scales larger than the free-streaming length of neutrinos of about 1 Gpc (for $m_\nu \sim 1$ eV) can thus be formed in a hot dark matter dominated universe.

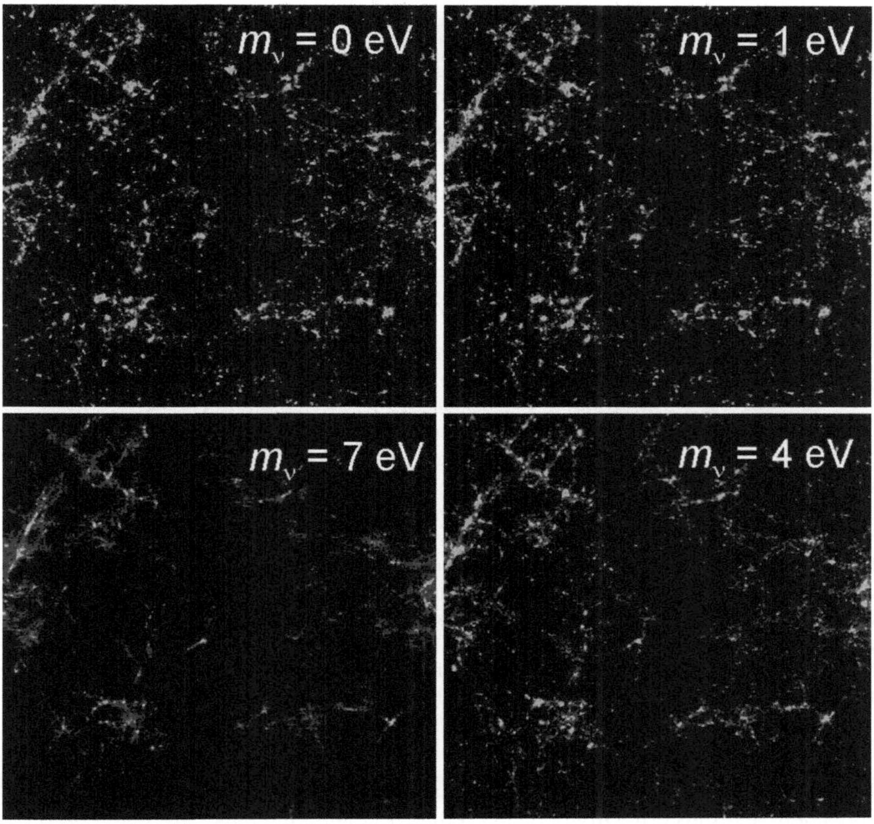

Fig. 2.9 Influence of the neutrino mass on small-scale structures of the universe. The plots show a simulation of the structure formation for different neutrino masses. For large neutrino masses the small scales are smeared out

The matter distribution of the universe can be measured by the Cosmic Microwave Background (CMB), galaxy and galaxy cluster surveys, weak lensing, Lyman-α-forest and 21-cm-line measurements [43]. The amount of small structures at different cosmological epochs is sensitive to the neutrino mass. It is evident that the more heavy the neutrino is, the more effective the washing out of small-scale structures will proceed. Figure 2.10 shows the power spectrum for different neutrino masses. Cosmology is sensitive to the absolute sum of neutrinos masses

$$m(\nu) = \sum_i m_{\nu_i}. \tag{2.37}$$

The current upper limits [43] are

$$m(\nu) < 0.5 - 2\,\text{eV}. \tag{2.38}$$

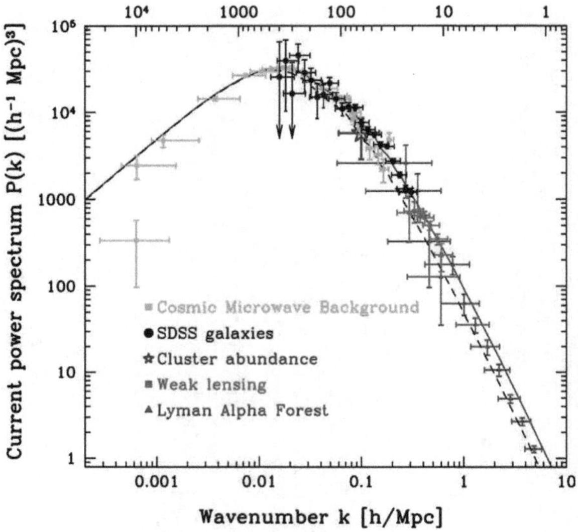

Fig. 2.10 Influence of the neutrino mass on the matter power spectrum. This plot shows the measurements and theoretical prediction of the matter power spectrum. The *solid line* represents a standard scenario, where the neutrino mass is assumed to be zero. The *dashed line* corresponds to where neutrinos contribute with 7 % to the dark matter in the universe, i.e. their mass is approximately 1 eV. The power spectrum is reduced by a factor of 2 for large wave numbers, i.e. small-scale structures [44]

The large variations are due to the fact that cosmological methods still are very model dependent and thus crucially depend on the parameter and data base being used.

2.4.2 Neutrinoless Double Beta Decay

A β-decay converting a neutron into a proton will transmute an even–even nucleus into an odd–odd nucleus, with generally less binding energy. In some cases, the daughter element with Z+1 protons has therefore a higher mass excess. As a consequence, the single β-decay is energetically forbidden and double β-decay as a second order weak process becomes experimentally observable (see Fig. 2.11). In this rare decay two electrons and two neutrinos are produced simultaneously

$$2n \rightarrow 2p + 2e^- + 2\overline{\nu}_e. \tag{2.39}$$

However, if the neutrino is its own antiparticle, i.e. a Majorana particle, the neutrino produced in one of the β-decays can be absorbed at the other vertex. In this case, there would be no neutrino in the final state. This process is called neutrinoless double β-decay ($0\nu\beta\beta$). In the Standard Model, which does not allow for lepton number violation, this decay is does not occur. Observing this decay would prove the Majorana nature of the neutrino.

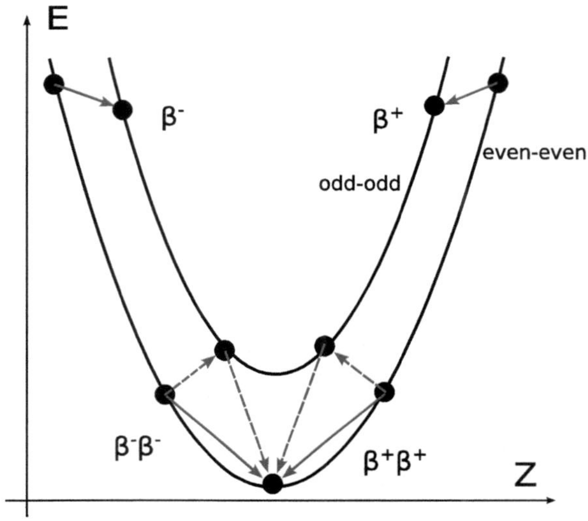

Fig. 2.11 Mass excess E as a function of number of protons Z. This figure illustrates that in some cases the single β-decay is energetically forbidden. This is the case for 60 naturally occuring isotopes. However, only in seven cases double β-decay has been experimentally observed

Fig. 2.12 Feynmangraph of neutrinoless double β-decay

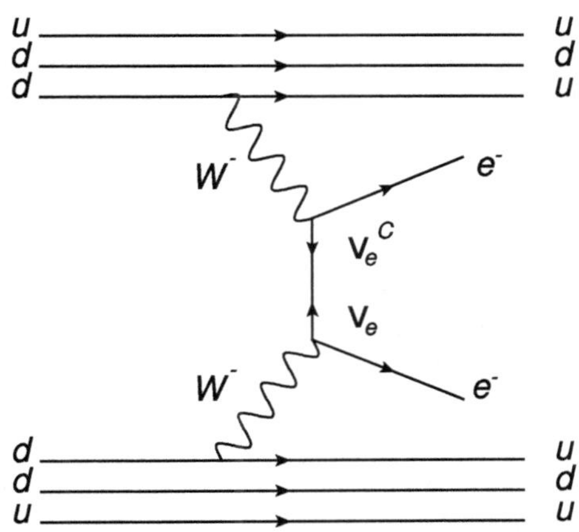

The decay amplitude Γ depends on the neutrino mass. To find the exact dependence, one considers the Feynman graph of $0\nu\beta\beta$ decay (see Fig. 2.12). The neutrino propagator is given by

$$\sum_i P_L U_{ei} \frac{\not{p} - m_i}{p^2 + m^2} U_{ei} P_L, \tag{2.40}$$

where $P_L = \frac{1-\gamma^5}{2}$ is a projector on a left chiral component of a field, accounting for the fact of maximal parity violation of the weak interaction and U is the PMNS matrix. As the matrices γ^5 and γ^μ anti-commute the $\not{p}$-term of the propagator vanishes. The mass parameter m^2 in the denominator can be neglected compared to p^2. One can then define an effective neutrino mass as

$$\Gamma \alpha \langle m\beta\beta \rangle = \left| \sum_i U_{ei}^2 m_{\nu_i} \right|. \tag{2.41}$$

with $\langle m\beta\beta \rangle$ denoting the Majorana neutrino mass.

It is important to note that $\langle m\beta\beta \rangle$ is a coherent sum of the neutrino masses. Since the PMNS matrix contains complex phases, cancellations are possible. As a result, even for relatively large m_{ν_i}, the contribution to $0\nu\beta\beta$ decay could be small. Additionally, the measurement is sensitive to uncertainties of the parameters of the PMNS matrix, especially the CP violating phases. When deducing the neutrino mass from the $0\nu\beta\beta$ decay rate one needs to take into account the fact that not only the massive Majorana neutrino contributes to the decay amplitude, but also any other corresponding physics beyond the Standard Model, e.g. right handed currents and SUSY, which can give rise to non-negligible contributions.

The experimental signature of $0\nu\beta\beta$ decay is a mono-energetic peak produced by the sum of the energies of the two electrons in the final state. The peak coincides with the endpoint of the continuous spectrum of the dominant $2\nu\beta\beta$ mode, as shown in Fig. 2.13. To deduce the neutrino mass from the experimental data, the lifetime $T_{1/2}$ of the $0\nu\beta\beta$ decay of a specific isotope is measured. The Majorana mass $\langle m\beta\beta \rangle$ depends on the lifetime $T_{1/2}$ and the nuclear matrix elements $|M_{nucl}|$

$$\langle m\beta\beta \rangle^2 = \left(T_{1/2} G(Q, Z) |M_{nucl}|^2 \right)^{-1} \tag{2.42}$$

where G(q,Z) denotes the phase space factor for $0\nu\beta\beta$ decay. The matrix elements can be computed with several methods, e.g. by a shell model ansatz, or a quasi particle radom phase approximation (QRPA), etc. However, the matrix elements of $0\nu\beta\beta$ for many nuclei are only know within an accuracy of about a factor of 2. The Heidelberg-Moscow experiment has claimed an experimental evidence for $\langle m\beta\beta \rangle \approx 0.4\,\text{eV}$ [45]. This result is disputed in the community and needs to be scrutinized by current and future experiments like the GERmanium Detector Array (GERDA) [46], CUORE [47], Super-Nemo [48], EXO [49], and MAJORANA [50].

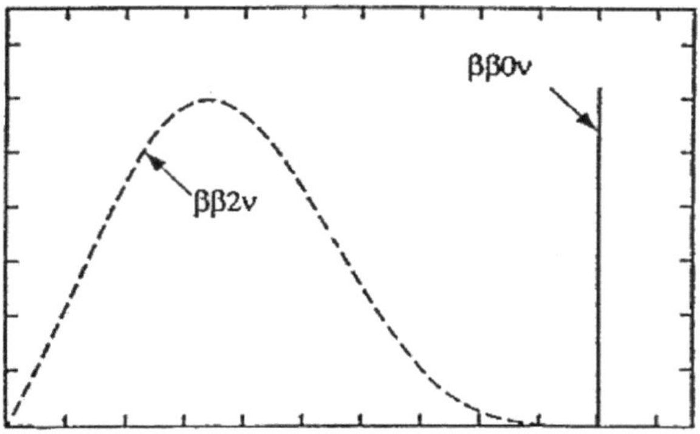

Fig. 2.13 Experimental signature of neutrinoless double β-decay. The figure shows the distribution of the sum of the electron energies ($T_1 + T_2$) normalized to the Q-value (mass difference of mother and granddaughter nucleus). The continuous spectrum is due to the $2\nu\beta\beta$ decay, the monoenergetic peak is the signature of $0\nu\beta\beta$ decay [51]

2.4.3 Single Beta Decay

The β^-decay of a nucleus is a weak decay in which a neutron is converted to a proton while an electron e^- and electron anti neutrino $\overline{\nu}_e$ are produced:

$$(A, Z) \rightarrow (A, Z+1) + e^- + \overline{\nu}_e \tag{2.43}$$

The kinematics of the β^--decay is sensitive to the neutrino mass. A nonzero neutrino mass reduces the endpoint energy and distorts the spectrum, especially in the vicinity of the endpoint, as visualized in Fig. 2.14. The β-electron energy spectrum is given by

$$\frac{dN}{dE_e} = C \cdot F(E, Z) \cdot p_e \cdot (E_e + m_e c^2) \cdot (E_0 - E_e) \sum_i |U_{ei}|^2 \sqrt{(E_0 - E_e)^2 - m_{\nu_i}^2}, \tag{2.44}$$

where F(E,Z) corresponds to the Fermi function, taking into account the Coulomb interaction of the outgoing electron with the daughter nucleus. The variables p_e, E_e and m_e are the electron momentum, kinetic energy and mass, respectively and E_0 is the endpoint energy for zero neutrino mass. Correspondingly, $E_0 - E_e$ is the total energy of the neutrino. The momentum of the neutrino is given by the sum of the momenta of each mass eigenstate, weighted by their fraction $|U_{ei}|^2$ within the electron flavor. The constant C is given by

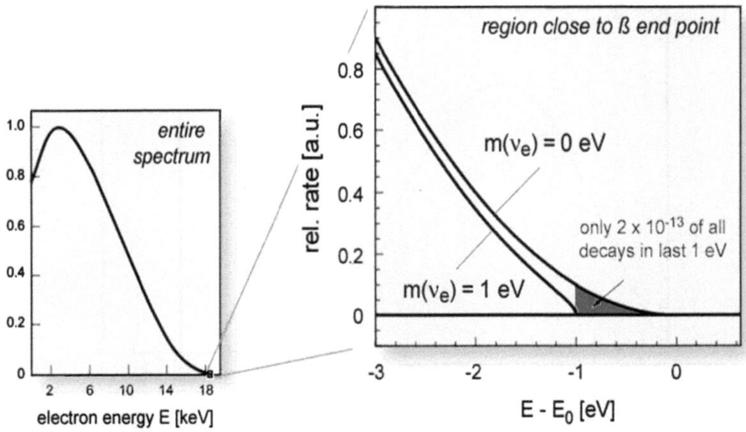

Fig. 2.14 Tritium β spectrum close to the endpoint. *Left* Full tritium spectrum. *Right* Zoom into the region close to the endpoint

$$C = \frac{G_F^2}{2\pi^3}\cos^2\Theta_C|M|^2. \tag{2.45}$$

with G_F being the Fermi constant and Θ_C the Cabbibo angle accounting for the transition strength of a down quark into an up quark (change of neutron into a proton). $|M|^2$ is the nuclear matrix element. In case of a super-allowed transition $|M|^2$ is not energy dependent close to the endpoint and therefore does not influence the shape of the spectrum.

For the region $(E_0 - E_e)^2 \ll m_{\nu_i}^2$ Eq. (2.44) can be expanded to

$$\frac{dN}{dE_e} \approx C \cdot F(E,Z) \cdot p_e(E_e + m_e c^2)(E_0 - E_e)^2 \sum_i |U_{ei}|^2 \left(1 - \frac{1}{2}\frac{m_{\nu_i}^2}{(E_0 - E_e)^2}\right) \tag{2.46}$$

$$= C \cdot F(E,Z) \cdot p_e(E_e + m_e c^2)\left((E_0 - E_e)^2 - \frac{1}{2}\sum_i |U_{ei}|^2 m_{\nu_i}^2\right) \tag{2.47}$$

$$= C \cdot F(E,Z) \cdot p_e(E_e + m_e c^2)(E_0 - E_e)\sqrt{(E_0 - E_e)^2 - \sum_i |U_{ei}|^2 m_{\nu_i}^2} \tag{2.48}$$

The expression $\sum_i |U_{ei}|^2 m_{\nu i}^2 =: \langle m_\beta \rangle^2$ is called the "effective electron antineutrino mass". In principle, a super-high resolution analysis of the spectrum would be sensitive to the single mass eigenstates. However, the experimental precision in β-spectroscopy does not allow to resolve the tiny mass splittings and therefore the effective electron neutrino mass $\langle m_\beta \rangle^2$ is measured. $\langle m_\beta \rangle^2$ is an incoherent sum of the neutrino masses, meaning that in contrast to the $0\nu\beta\beta$ method no cancellations can occur. Furthermore, the single β-decay is not sensitive to the intrinsic nature of the neutrino, i.e. whether it is a Majorana or Dirac particle.

In a measurement of the β-spectrum all parameters that modify the spectrum, such as electronic final states, recoil energy of the daughter nucleus and radiative corrections need to be taken into account. To be sensitive to neutrino masses in the cosmologically preferred sub-eV range, one needs an instrument with a very good energy resolution, high luminosity and very low background. The current experimental limit from analysis of tritium β-decay is $\langle m_\beta \rangle < 2\,\text{eV}$ from the Mainz and Troizk experiments [8–11].

These experiments have pioneered high precision β-spectroscopy of molecular tritium with the so-called MAC-E filter technique [52, 53]. However, due to limitations in the experimental resolution ΔE, and, more importantly, due to limits of the source intensity, the sensitivity of both experiments did not allow to explore the sub-eV range. In the following chapter the design principles and criteria of the next-generation tritium β-decay experiment KATRIN will be presented, which allows to push the precision in β-spectroscopy to unprecedented levels.

References

1. R. Davis, A review of the homestake solar neutrino experiment. Prog. Part. Nucl. Phys. **32**, 13–32 (1994)
2. B. Cleveland, T. Daily, R.D. Jr., J. Distel, K. Lande, C. Lee, P. Wildenhain, J. Ullman, Update on the measurement of the solar neutrino flux with the homestake chlorine detector. Nucl. Phys. B Proc. Suppl. **38**(1–3), 47–53 (1995). Neutrino 94
3. Y. Fukuda et al., Evidence for oscillation of atmospheric neutrinos. Phys. Rev. Lett. **81**, 1562–1567 (1998)
4. Q.R. Ahmad, Direct evidence for neutrino flavor transformation from neutral-current interactions in the sudbury neutrino observatory. Phys. Rev. Lett. **89**, 011301 (2002)
5. Q.R. Ahmad, Measurement of the rate of $\nu_e + d \rightarrow p + p + e^-$ interactions produced by $B8$ solar neutrinos at the sudbury neutrino observatory. Phys. Rev. Lett. **87**, 071301 (2001)
6. K. Abe et al., Indication of electron neutrino appearance from an accelerator-produced off-axis muon neutrino beam. Phys. Rev. Lett. **107**, 041801 (2011)
7. H.D. Kerrect, First results from the Double Chooz experiment, in *LowNu 2011* (2011)
8. C. Kraus, B. Bornschein, L. Bornschein, J. Bonn, B. Flatt, A. Kovalik, B. Ostrick, E.W. Otten, J.P. Schall, T. Thümmler, C. Weinheimer, Final results from phase II of the Mainz neutrino mass search in tritium. Eur. Phys. J. C **40**, 447–468 (2005)
9. J. Bonn, B. Bornschein, L. Bornschein, L. Fickinger, B. Flatt, O. Kazachenko, A. Kovalik, C. Kraus, E.W. Otten, J.P. Schall, H. Ulrich, C. Weinheimer, The mainz neutrino mass experiment. Nucl. Phys. B Proc. Suppl. **91**(1–3), 273–279 (2001). Neutrino 2000
10. V.M. Lobashev, V.N. Aseev, A.I. Belesev, A.I. Berlev, E.V. Geraskin, A.A. Golubev, O.V. Kazachenko, Y.E. Kuznetsov, R.P. Ostroumov, L.A. Rivkis, B.E. Stern, N.A. Titov, S.V. Zadorozhny, Y.I. Zakharov, Direct search for mass of neutrino and anomaly in the tritium beta-spectrum. Phys. Lett. B **460**(1–2), 227–235 (1999)
11. V.M. Lobashev, V.N. Aseev, A.I. Belesev, A.I. Berlev, E.V. Geraskin, A.A. Golubev, O.V. Kazachenko, Y.E. Kuznetsov, R.P. Ostroumov, L.A. Rivkis, B.E. Stern, N.A. Titov,

C.V. Zadoroghny, Y.I. Zakharov, Direct search for neutrino mass and anomaly in the tritium beta-spectrum: Status of troitsk neutrino mass experiment. Nuclear Physics B - Proceedings Supplements **91**(1–3), 280–286 (2001). Neutrino 2000
12. J. Chadwick, Intensitätsverteilung im magnetischen Spektrum von β-Strahlen von Radium B+C. Verh. d. Deutsch. Phys. Ges. **15**, 383 (1914)
13. W. Pauli, Letter to Gauvereinstagung in Tübingen: "Sehr geehrte radioaktive Damen und Herren", published in R. Kronig and V. Weisskopf (Eds.), Wolfgang Pauli, Collected scientivic Papers, Vol. 2, Interscience, New York (1964), 1930.
14. E. Fermi, Versuch einer Theorie der β-Strahlen. Z. Phys. A: Hadrons Nucl. **88**, 161–177 (1934). doi:10.1007/BF01351864
15. F.A. Scott, Energy spectrum of the β-rays of radium E. Phys. Rev. **4** (1935)
16. C.L. Cowan, F. Reines, F.B. Harrison, H.W. Kruse, A.D. McGuire, Detection of the free neutrino: a confirmation. Science **124**(3212), 103–104 (1956)
17. F. Reines, C.L. Cowan, F.B. Harrison, A.D. McGuire, H.W. Kruse, Detection of the free antineutrino. Phys. Rev. **117**, 159–173 (1960)
18. J.-M. Gaillard, K. Goulianos, L.M. Lederman, N. Mistry, M. Schwartz, J. Steinberger, G. Danby, Observation of high-energy neutrino reactions and the existence of two kinds of neutrinos. Phys. Rev. Lett. **9**, 36–44 (1962)
19. K. Kodama, Observation of tau neutrino interactions. Phys. Lett. B **504**(3), 218–224 (2001)
20. D. DeCamp et al., A precise determination of the number of families with light neutrinos and of the Z boson partial widths. Phys. Lett. B **235**(3–4), 399–411 (1990)
21. D. DeCamp et al., Determination of the number of light neutrino species. Phys. Lett. B **231**(4), 519–529 (1989)
22. J.N. Bahcall, A.M. Serenelli, S. Basu, New solar opacities, abundances, helioseismology, and neutrino fluxes. Astrophys. J. Lett. **621**(1), L85 (2005)
23. Borexino homepage. http://borex.lngs.infn.it/. Last update: 01 Jul 2010
24. G. Bellini et al., Precision measurement of the ^{7Be} solar neutrino interaction rate in borexino. Phys. Rev. Lett. **107**, 141302 (2011)
25. Homepage of J. Bahcall. http://www.sns.ias.edu/ jnb/. Last update: 2005
26. B. Aharmim et al., Determination of the ν_e and total $B8$ solar neutrino fluxes using the sudbury neutrino observatory phase I data set. Phys. Rev. C **75**, 045502 (2007)
27. B. Kayser, B-meson and neutrino oscillation: a unified treatment. arXiv:1110.3047v1 [hep-ph]
28. C. Giunti, C.W. Kim, J.A. Lee, U.W. Lee, On the treatment of neutrino oscillations without resort to weak eigenstates. Phys. Rev. **D48**, 4310–4317 (1993)
29. C. Giunti, Neutrino wave packets in quantum field theory. JHEP **11**, 017 (2002)
30. S. Abe et al., Precision measurement of neutrino oscillation parameters with KamLAND. Phys. Rev. Lett. **100**, 221803 (2008)
31. K. Nakamura, P.D. Group, Review of particle physics. J. Phys. G: Nucl. Part. Phys. **37**(7A), 075021 (2010)
32. L. Wolfenstein, Neutrino oscillations in matter. Phys. Rev. D **17**, 2369–2374 (1978)
33. E. Majorana, Teoria simmetrica dell elettrone e del positrone. Il Nuovo Cimento (1924–1942) **14**, 171–184 (1937). doi:10.1007/BF02961314
34. S. Antusch, M. Drees, J. Kersten, M. Lindner, M. Ratz, Neutrino mass operator renormalization in two Higgs doublet models and the MSSM. Phys. Lett. B **525**(1–2), 130–134 (2002)
35. N. Arkani-Hamed, S. Dimopoulos, G. Dvali, J. March-Russell, Neutrino masses from large extra dimensions. Phys. Rev. D **65**, 024032 (2001)
36. J. Casas, J. Espinosa, A. Ibarra, I. Navarro, Nearly degenerate neutrinos, supersymmetry and radiative corrections. Nucl. Phys. B **569**(1–3), 82–106 (2000)
37. E. Ma, Verifiable radiative seesaw mechanism of neutrino mass and dark matter. Phys. Rev. D **73**, 077301 (2006)
38. M. Gell-Mann, P. Ramond, R. Slansky, Complex spinors and unified theories. Prepared for Supergravity Workshop (Stony Brook, New York, 1979)
39. R.N. Mohapatra, G. Senjanović, Neutrino masses and mixings in gauge models with spontaneous parity violation. Phys. Rev. D **23**, 165–180 (1981)

40. G. Altarelli, K. Winter, *Neutrino Mass*. Springer Tracts in Modern Physics (Springer, 2003)
41. R.N. Mohapatra, Theory of neutrinos: a white paper. Rep. Prog. Phys. **70**(11), 1757 (2007)
42. K. Collaboration, KATRIN Design Report (FZKA Report 7090), Tech. Rep., KIT (2004)
43. S. Hannestad, Neutrino physics from precision cosmology. Prog. Part. Nucl. Phys. **65**(2), 185–208 (2010)
44. M. Tegmark, Cosmological neutrino bounds for non-cosmologists. Phys. Scr. **2005**(T121), 153 (2005)
45. H.V. Klapdor-Kleingrothaus, A. Dietz, H.L. Harney, I.V. Krivosheina, Evidence for neutrinoless double beta decay. Mod. Phys. Lett. **A16**, 2409–2420 (2001)
46. GERDA homepage. http://www.mpi-hd.mpg.de/gerda/. Last update: 04 Nov 2010
47. M. Sisti, The Cuore collaboration, From cuoricino to CUORE: investigating neutrino properties with double beta decay. J. Phys.: Conf. Ser. **203**(1), 012069 (2010)
48. Super-Nemo homepage. http://nemo.in2p3.fr/supernemo/. 30 Oct 2006
49. Exo-200 homepage. http://www-project.slac.stanford.edu/exo/. 12 Sep 2011
50. Majorana homepage. http://majorana.npl.washington.edu/. 04 June 2009
51. R. Beuselinck, J. Sedgbeer, Y. Shitov, The supernemo experiment. http://www.imperial.ac.uk/research/hep/research/supernemo.htm. Last update: 2011
52. V.M. Lobashev, P.E. Spivak, A method for measuring the electron antineutrino rest mass. Nucl. Instrum. Methods Phys. Res., Sect. A **240**(2), 305–310 (1985)
53. A. Picard, H. Backe, H. Barth, J. Bonn, B. Degen, T. Edling, R. Haid, A. Hermanni, P. Leiderer, T. Loeken, A. Molz, R. Moore, A. Osipowicz, E. Otten, M. Przyrembel, M. Schrader, M. Steininger, C. Weinheimer, A solenoid retarding spectrometer with high resolution and transmission for keV electrons. Nucl. Instrum. Methods Phys. Res., Sect. B **63**(3), 345–358 (1992)

Chapter 3
Introduction to the KATRIN Experiment

The aim of the Karlsruhe Tritium Neutrino (KATRIN) experiment is the measurement of the absolute scale of the effective electron antineutrino mass m_β by high precision spectroscopy of the β-decay of molecular tritium close to the endpoint E_0.

To explore the sub-eV region of neutrino masses three different experimental techniques are currently pursued and investigated. The most established technique is the MAC-E-Filter principle. In that case an integral energy spectrum close to the endpoint (where the influence of the neutrino mass is maximal) is measured by using an electrostatic filter. KATRIN, which is currently under construction at the Karlsruhe Institute of Technology (KIT), takes advantage of this approach and will push this technology to its limits. More details will be explained in the following sections. Another generic possibility is to measure the energy of the β-decay electrons by recording the minute temperature change in the β-substrate with micro-bolometers. In this case the full differential spectrum is measured, and the source can be used as detector at the same time. The MARE-2 experiment in Italy is currently constructing an array of low temperature calorimeters to measure Rhenium β-decay [1]. Finally, the energy of β-electrons can be measured by detecting their synchrotron radiation. For that purpose the electrons are stored in a magnetic trap and antennas surrounding the trap record the radiation. Project 8 [2] is currently investigating the feasibility of this approach.

In Sect. 3.1, an overview of the components of the KATRIN experiment will be given. Section 3.2 addresses the working principle of the spectrometers. In Sect. 3.3 statistical and systematic uncertainties of KATRIN will be discussed and an overview of background sources is given in Sect. 3.4.

3.1 Experimental Overview

The basic idea of the KATRIN experiment is to implement a molecular tritium source of the highest stability and luminosity in combination with a variable retarding potential. This filter transmits only those electrons which have more energy than the filter

S. Mertens, *Background Processes in the Electrostatic Spectrometers of the KATRIN Experiment*, Springer Theses, DOI: 10.1007/978-3-319-01177-6_3,

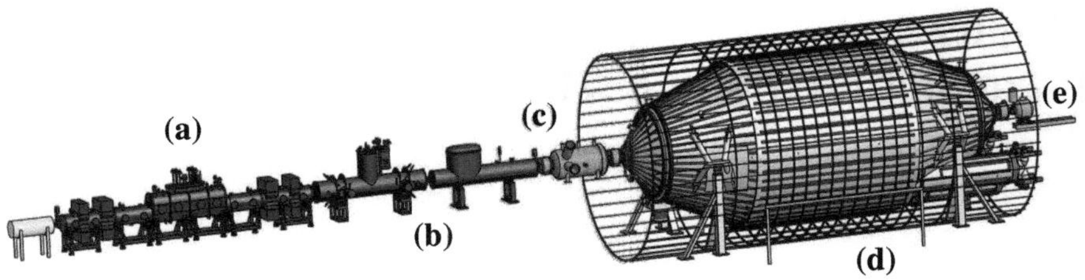

Fig. 3.1 **a** Windowless gaseous tritium source: β-decay of molecular tritium. **b** Transport section: reduction of tritium number density and adiabatic guidance of electrons to the spectrometers. **c** Pre-spectrometer: pre-filter of β-spectrum. **d** Main spectrometer: energy analysis of β-electrons. **e** Detector: measurement of transmitted electrons [3]

voltage to a detector for counting. By measuring the count rate for different retarding voltages the shape of the integrated energy spectrum can be determined. Figure 3.1 shows a schematic view of the KATRIN setup. In the following the subtleties and challenges associated with each component will be described.

3.1.1 Tritium Source

The hydrogen isotope tritium offers several unique advantages as β-electron emitter

$$^3_1\mathrm{H} \rightarrow ^3_2\mathrm{He}^+ + e^- + \overline{\nu}_e \tag{3.1}$$

- Tritium has a very low endpoint energy of $E_0 = 18.6\,\mathrm{keV}$. On the one hand the total count rate strongly increases with the endpoint energy $\Gamma \propto E_0^4$ (for low E_0) $\Gamma \propto E_0^5$ (for high E_0), on the other hand the relative fraction of electrons in the interesting energy region close to the endpoint decreases with E_0 ($\propto E_0^{-3}$). Correspondingly, the total count rate in the last 10 eV below the endpoint only weakly depends on the endpoint energy E_0. A small value of the endpoint energy, however, still has the advantage of technical feasibilty with regard to electro-static filters.
- The $^3\mathrm{H} \rightarrow ^3\mathrm{He}$ transition is a superallowed β^--decay. As a consequence, the lifetime of $T_{1/2} = 12.3$ a is rather short. Correspondingly, high count rates can be achieved with rather low source densities. Furthermore, the nuclear matrix element is energy independent and easily computable.
- Molecular tritium can be used in gaseous form at low temperatures. The gaseous form is preferred compared to a condensed source due to lower systematic uncertainties and the feasibility of higher count rates.
- Tritium has a low Z value. Consequently, the electronic final states can be computed rather easily using the "sudden approximation" approach. Moreover, the Fermi function F(Z = 2, E) of $^3_2$He is almost energy independent close to the endpoint

As tritium has to be used in molecular form at low temperatures, vibrational and rotational excitations of the daughter molecule ^{3}He need to be taken into account.

Technical Implementation of the Windowless Gaseous Tritium Source

In the KATRIN experiment, tritium of very high isotopic purity ($> 95\,\%$) is injected through capillaries into the 10 m long Windowless Gaseous Tritium Source (WGTS) tube. The T_2 molecules then diffuse over a distance of 5 m to both ends of the WGTS. The WGTS beam tube is situated in a magnetic field of $B_{\text{source}} = 3.6$ T, which is oriented in beam direction. All β-electrons that are emitted in forward direction are guided along the field lines towards the spectrometers.

The source tube, and the tritium molecules therein, are kept at a very low temperature of $T = 27\,$K. On the one hand, ultra-cold molecular tritium allows to achieve a high column density $\rho d = 5 \cdot 10^{17}$ cm^{-2} with moderate injection rates of $5 \cdot 10^{19}$ molecules/s. Furthermore, ultra-cold tritium has only a small molecular motion, resulting in a low Doppler-shift of the β-electron energy. Finally, undesired plasma effects of the source are suppressed for low temperatures.

The column density ρd of $5 \cdot 10^{17}$ molecules/cm^2 defines the source luminosity and has to be kept very stable. The degree of stability of the column density is the main systematic uncertainty of the WGTS. It is important to guarantee a constant activity, and secondly, to maintain a time independent ionization probability of the signal electrons. To achieve a stable column density both the injection pressure and the temperature of the 10 m long beam tube must be very stable. For this pupose a two phase Neon cooling system is used. Actual results of temperature stability test measurements with the so called WGTS-demonstrator show that the temperature variations are much smaller than 30 mK as desired [4]. Figure 3.2 shows a schematic view of the WGTS cryostat, which is currently under construction.

3.1.2 Transport Section

The task of the transport section is to guide the β-decay electrons adiabatically from the WGTS to the spectrometers. Since the spectrometer section must be essentially tritium free, the tritium flow must be reduced from the injection rate of 1.8 mbar $\cdot\, \ell$/s to 10^{-14} mbar $\cdot\, \ell$/s at the end of the transport section. This unprecedented large suppression factor will be achieved by a combination of differential and cyrogenic pumping.

Differential Pumping Section

Modules for differential pumping are installed both at the rear section of the WGTS (DPS1-R) and at the front side (DPS1-F, DPS2-F). To avoid the beaming effect of the neutral tritium molecules, the beam lines DPS2-F cyrostat is designed with chicanes

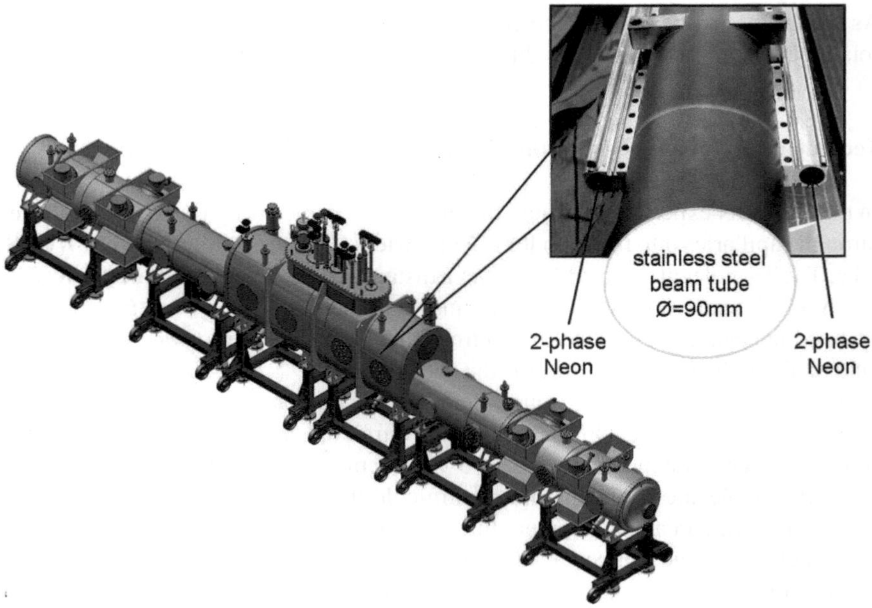

Fig. 3.2 The figure shows the 16 m long cyrostat in which the beam tube of 90 mm diameter with the two phase liquid neon cooling system is situated [5]

of 20°. The magnetic field guides the electrons through this system, whereas the tritium molecules collide with the walls of the beam tube. To make use of this effect, the DPS2-F is equipped with a series of four pump ports, in which turbomolecular pumps (TMPs) with very high pumping capacity are installed (see Fig. 3.3). The combined active pumping reduces the tritium flow by an estimated factor of about 10^7.

Decay product of T_2 is the positive molecular ion ${}^3_2HeT^+$. As in the case for the β-electrons, the ions are guided along the magnetic field lines around the chicanes. Therefore, pumping by TMPs is not efficient for ions. For this reason, an electrode with a slightly more positive potential is implemented at the end of the DPS2-F to restrain the positive ions from entering the CPS. As a results they are reflected and fly back towards the WGTS. However, there they are again reflected by the gas pressure. To remove the trapped ions, an electric dipole electrode has to be installed in the DPS2-F. The electric field together with magnetic field creates an $\vec{E} \times \vec{B}$-drift that eventually pushes the ions to the wall [6].

At the entrance and exit of the DPS2-F beam line two FT-ICR (Fourier Transform - Ion Cyclotron Resonance) modules are installed to measure the number density and composition of ions. There, ions are trapped in a penning trap for a short time and their cyclotron frequency and therefore their mass-to-charge ratios are measured by the FT-ICR technique [7]. This is of great importance for the experiment, as a variety of different ions is produced via following processes: Excited ${}^3_2HeT^+$, that

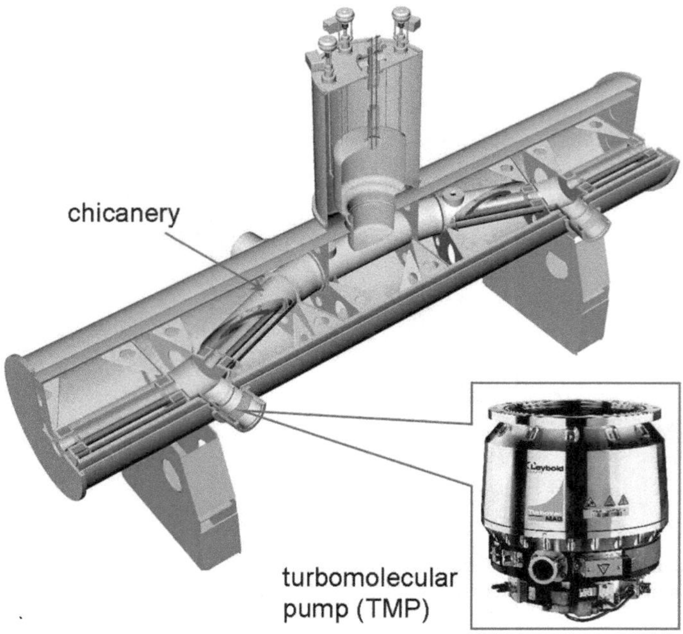

Fig. 3.3 The figure shows the 6.96 m long DPS2-F, which is situated after the WGTS. The beamtube, that guides the β-electrons has a chicanery shape to remove the non decayed tritium molecules. At the two pumports turbomolecular pumps are installed. At the end of the DPS the ion trap will be mounted [5]

is produced in the β-decay can dissociate into ${}^3_2\mathrm{He}^+$ and T^+. Moreover, T_2 can be ionized to T_2^+, which subsequently produces T_3^+, T_5^+, when interacting with another T_2 [8].

Cyrogenic Pumping Section

The inner surface of the cyrogenic pumping section (CPS) contains beam tube elements at 3 K which are covered with argon snow to passively cryosorb tritium molecules. Any tritiated molecule hitting the argon frost surface is cryosorbed and permanently fixed, as visualized in Fig. 3.4. The CPS will accumulate about 10^{17} tritium molecules per day and is expected to further reduce the tritium flow by a factor of 10^7.

The CPS is currently being manufactured by an industrial partner in Genoa (Italy) and will be shipped to KIT in the near future.

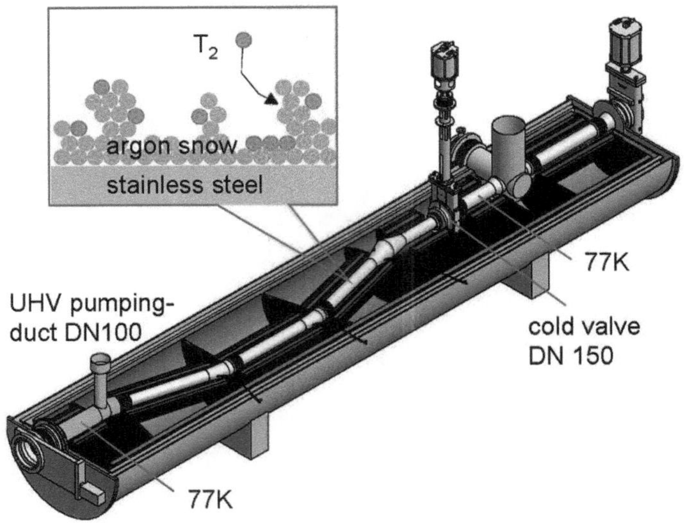

Fig. 3.4 The figure shows CPS, which is situated after the DPS2-F. The inner surface is covered with argon snow to capture the remaining tritium molecules [5]

3.1.3 Spectrometer Section

The spectrometer section consists of two electrostatic retarding filters: the pre-spectrometer and the main spectrometer. The pre-spectrometer offers the option to act as a pre-filter, reflecting all electrons 300 eV below the endpoint, while transmitting the interesting part of the spectrum undisturbed. All electrons transmitted through this first stage are guided to the main spectrometer for precise energy analysis with an energy resolution of $\Delta E = 0.93$ eV. Both spectrometer are of the MAC-E-Filter type (see also Sect. 3.2). A third spectrometer, the monitor spectrometer, is used to monitor the high voltage of the main spectrometer.

Pre-spectrometer

One of the pre-spectrometer's major tasks has been to serve as a prototype for advanced technologies and experimental methods later applied to the much larger main spectrometer. Many basic concepts, such as the design of the ultra high vacuum (UHV) system and high voltage stabilization were successfully tested at the pre-spectrometer. Especially important was the detailed investigation of background processes in MAC-E-filters. As explained in more detail in Chaps. 6 and 7 two new classes of background were identified and studied at the pre-spectrometer. The knowledge gained during the extensive measurement period at the pre-spectrometer has proven to be extremely valuable for the design and operation of the main spectrometer.

In the reference KATRIN setup the pre-spectrometer was initially designed to pre-filter the β-electrons, thereby reducing the flux by six orders of magnitude. To do so, the pre-spectrometer works as an electrostatic filter set to a fixed retarding voltage at 300 V below the endpoint. At this operating point, the energy resolution of the pre-spectrometer of $\Delta E = 70$ eV is sufficient to leave the interesting high-energy part of the β-spectrum unmodified. The reason to aim for reducing the flux was to avoid the creation of background. The main concern was that β-electrons penetrating the main spectrometer could scatter on residual gas molecules there and ionize them. The resulting positive ions are not reflected by the negative potential of the main spectrometer. If they in turn ionize residual gas the resulting secondary electrons can reach the detector and produce an irreducible background. On the other hand, running two spectrometers next to each other produces a large electric potential trap for electrons leading potentially to an even higher background (see Sect. 3.4).

As will be shown in Chap. 7, it was found in this work that the background produced by tritium decays in the volume of the main spectrometer was underestimated. The partial pressure of tritium has to be reduced by another 2 orders of magnitude in the main spectrometer. Therefore, the pre-spectrometer might take over the role of a further tritium pump. Accordingly, it is planned to equip the pre-spectrometer with up to 1 km getter material to absorb the tritiated molecules before they can enter the main spectrometer.

The pre-spectrometer has a length of 3.4 m and a diameter of 1.7 m. At both ends a superconducting magnet is installed providing a magnetic field of 4.5 T at the center of the magnet and 15.6 mT in the center of the spectrometer. At the entrance of the spectrometer an e-gun is installed to test the transmission properties [9]. At the end a 64 Silicon PIN-diode detector is mounted, which allows for spacial resolution. As a novel design feature, if compared to the Mainz and Troizk set-ups, the tank itself is set on high voltage. An inner electrode system consisting of two full electrodes and a wire electrode can be set to a different potential as the tank. Figure 3.5 shows a schematic view of the pre-spectrometer. The vacuum system of the pre-spectrometer is described in detail in Appendix F.

Main Spectrometer

The exceedingly large dimensions of the main spectrometer allow to operate it as an extremely precise high energy filter, operated on a stable high voltage in the ppm range. The scanning voltage is varied in steps of $\Delta U = 0.5 - 1$ V around a narrow region close to the endpoint at $E_0 = 18.6$ kV. The highest electrostatic potential is located in the central plane of the spectrometer, perpendicular to the beam axis, the so called "analyzing plane".

The β-electrons from the WGTS are guided along the magnetic field lines, towards the spectrometers. As a result of their isotropic emission at the source, they follow a cyclotron motion along the field lines. To achieve a very high energy resolution,

Fig. 3.5 [5]. *1* superconducting solenoids, *2* pre-spectrometer vessel, *3* inner electrode system, *4* 90° pump port, *5* 45° pump port, *6* insulator

the cyclotron motion needs to be almost fully transformed into motion parallel to the magnetic field lines, since only the latter helps them to overcome the retarding potential. To obtain adequate parallelization, the magnetic field has to drop by four orders of magnitude from the entrance to the center of the main spectrometer. Since the magnetic flux Φ is conserved, the cross section of the flux tube in the center plane has to be four orders of magnitude larger than at the entrance. This scaling explains the huge size of the main spectrometer (Length L = 23.8 m, diameter d = 9.8 m, cross sectional area A = 650 m^2 and volume V = 1,400 m^3) (see Fig. 3.7). The subtle interplay of the electric potential rise with the magnetic field decrease is visualized in Fig. 3.6.

To compensate the distorting earth magnetic field and to fine-shape the magnetic field inside the main spectrometer, the vessel is surrounded by a huge air coil system (see Fig. 3.8). It consists of 10 horizontal current loops and 16 vertical ones to compensate the earth magnetic field (EMCS). Additionally, a system of 15 Helmholtz-like coils with individually adjustable currents allows for precise adjusting of the gradients and the overall strength of the magnetic field (LFCS). The magnetic design of the LFCS was part of this thesis (for details see Appendix A).

Inside the main spectrometer an inner electrode system has been installed. A ground electrode at the entrance (exit) assures a smooth rise (fall) of the potential. A two layer wire electrode (see Fig. 3.8) with more than 24,000 wires of a diameter of 300 μm (outer layer) and 200 μm (inner layer) at a distance of 15 cm (outer layer) and 22 cm (inner layer) to the surface of the main spectrometer tank, serves as additional shielding against muon induced electrons from the wall (see Chap. 5) as well as potential defining element.

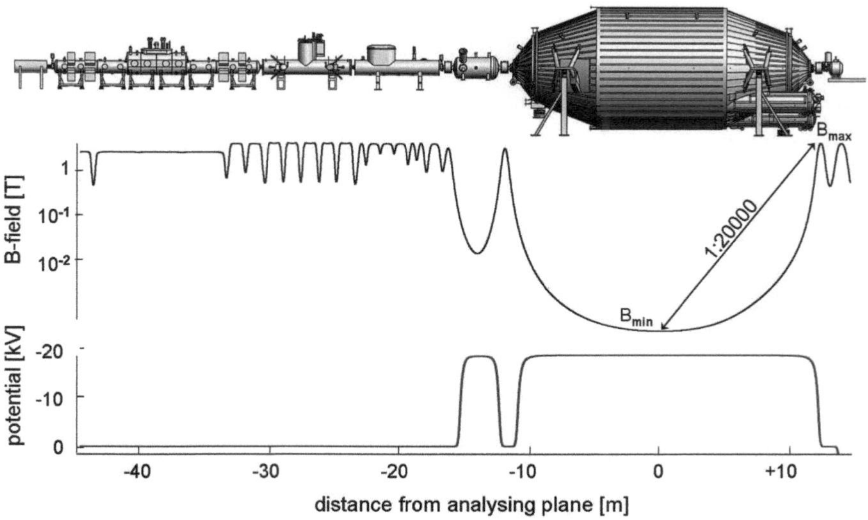

Fig. 3.6 This figure shows the interplay of the electric potential and the magnetic field throughout the whole KATRIN setup. One can see the retarding potential applied at the two spectrometers and the corresponding magnetic field distribution [5]

To achieve a pressure in the UHV regime of 10^{-11} mbar the main spectrometer is equipped with six cascaded Turbo Molecular Pumps (TMPs) connected to the main spectrometer vessel in parallel, and three large getter pumps with an overall pumping speed of 10^6 ℓ/s for H_2. More details can be found in Appendix F.

Monitor Spectrometer

The former Mainz spectrometer is being integrated in the KATRIN setup as precise high voltage monitor. It is operated as a high-resolution MAC-E-Filter coupled with a krypton source as a nuclear standard. The voltage of the monitor spectrometer is directly fed by the high voltage on the main spectrometer. By scanning the mono-energetic, narrow 17.8 keV ^{83m}Kr (K-32) line, variations of the retarding potential on the ppm scale can be observed. Hence, the stability of the retarding potential is continuously being monitored. An independent cross-check is provided by a precise high-voltage divider coupled to a precision voltmeter [11].

3.1.4 Focal Plane Detector

The focal plane detector is a semi-conductor based silicon PIN diode. Its main goal is to detect transmitted electrons with a detection efficiency of $> 90\,\%$.

Fig. 3.7 The main spectrometer was maneuvered through Leopoldshafen on November 2006. It has been manufactured by MAN DWE at the Deggendorf plant, however, due its size it could not be transported to Karlsruhe on motorways. Therefore, rather than taking a journey of about 400 km, the spectrometer had to travel nearly 9,000 km through the Danube River, the Black Sea, the Mediterranean Sea, the Atlantic Ocean and the River Rhine [5, 10]

Fig. 3.8 *Left* The air coil system surrounds the main spectrometer vessel. Its purpose is to compensate the earth magnetic field and to fine tune the magnetic field inside the main spectrometer. *Right* View into the main spectrometer. Its inner surface is covered with 24,000 wires of 200–300 μm thickness, providing additional shielding against muon-induced background as well as fine tuning the electrostatic retarding voltage

The detector is subdivided into 148 pixels to achieve good spatial resolution (see Fig. 3.9). This is important, as electrons passing the analyzing plane at different radii will experience slightly different retarding potentials, due to the radial inhomogeneity of the electric potential. To account for this effect the detector consists of 12 concentric rings subdivided azimuthally into 12 pixels each. This allows for a

Fig. 3.9 The electrons are guided along magnetic field lines to the sensitve area of the detector, which has a diameter of 10 cm. The inset shows a photo of the detector, in which the 148 pixels are visible [5]

precise mapping of inhomogeneities of the retarding potential. Each detector pixel measures an independent tritium β-spectrum, which has to be corrected for the actual retarding potential.

The detector is situated in a superconducting magnet providing a magnetic field of $B_{\text{det}} = 3 - 6$ T. The detector magnet is adjacent to the so called pinch magnet which provides the maximal magnetic field of $B_{\text{max}} = 6$ T of the entire KATRIN setup. All electrons that started in the source with an angle larger than 51° will be reflected by the pinch magnet before they reach the detector. This is advantageous, as electrons emitted under a large angle, perform a lot of cyclotron motion, which in turn increases their total path length and therefore their scattering probability. To exclude those electrons the magnetic field of the source is not the maximal field.

There is an option to use post-acceleration to increase the kinetic energy of the β-electrons to about 30 keV or above before they hit the detector. This could help in discriminating signal from fluorescence background. Additionally, the impact angle of the electrons relative to the detector surface would be increased, which decreases the probability of backscattering from the detector surface [12].

3.2 MAC-E-Filter Principle

About 10^{11} β-electrons are produced isotropically in the windowless gaseous tritium source (WGTS). However only in about 10^{-13} of the decays, an electron in the interesting energy region (i.e. 1 eV below the endpoint) is produced. To analyze as many electrons as possible, a sizable fraction of the forward half cone of the

β-electrons is guided along magnetic field lines to the spectrometers. The momentum of the electrons is composed of a component parallel (with longitudinal energy $E_{\parallel}$) and transversal (with cyclotron energy $E_{\perp}$) to the magnetic field lines. The kinetic energy can thus be written as

$$E_{\text{kin}} = E_{\perp} + E_{\parallel}. \tag{3.2}$$

However, only the longitudinal part $E_{\parallel}$ is analyzed by the electrostatic filter. To achieve both high count rates and good energy resolution the $E_{\perp}$ component needs to be transformed almost entirely into $E_{\parallel}$ on the way to the analyzing plane. This is achieved by the MAC-E-Filter principle, where the magnetic field drops by four orders of magnitude from the source ($\left|\overrightarrow{\mathrm{B}}\right|_{\mathrm{s}} = 3.6\,\mathrm{T}$) to the analyzing plane ($\left|\overrightarrow{\mathrm{B}}\right|_{\mathrm{a}} = 3 \cdot 10^{-4}$ T). The reduction of the magnetic field proceeds over a length of around 10 m in the main spectrometer, assuring a smooth change and therefore an adiabatic motion of the electrons. For electrons moving adiabatically the orbital magnetic moment μ is conserved. In first order, μ is given by

$$\mu \approx \frac{E_{\perp}}{\left|\overrightarrow{\mathrm{B}}\right|} \approx \text{const.} \tag{3.3}$$

The transversal energy of the electron in the center of the spectrometer can thus be expressed as

$$E_{\perp}^{\mathrm{a}} = E_{\perp}^{\mathrm{s}} \frac{\left|\overrightarrow{\mathrm{B}}\right|^{\mathrm{a}}}{\left|\overrightarrow{\mathrm{B}}\right|^{\mathrm{s}}} = \frac{E_{\perp}^{\mathrm{s}}}{12,000}, \tag{3.4}$$

for B_{s} and B_{a} given above. The index indicates the position (s = source and a = analyzing plane). The cyclotron energy in the analyzing plane $E_{\perp}^{\mathrm{a}}$ is, as the magnetic field, reduced by four orders of magnitude compared to the cyclotron energy in the source $E_{\perp}^{\mathrm{s}}$. On the way to the analyzing plane $E_{\perp}$ is almost completely transformed into $E_{\parallel}$, which is then analyzed by the electrostatic filter. When the electron adiabatically moves from low to high magnetic field its longitudinal energy is transformed back into the original transversal energy.

As mentioned, the maximal magnetic field is applied not at the source but at the pinch magnet close to the detector. This implies that electrons starting with an angle larger than $\theta^{\max}$ are reflected by the maximal magnetic field. To calculate the maximal acceptance angle $\theta^{\max}$ one considers

$$\frac{E_{\perp}^{\mathrm{s}}}{\left|\overrightarrow{\mathrm{B}}\right|^{\mathrm{s}}} = \frac{E_{\perp}^{\mathrm{p}}}{\left|\overrightarrow{\mathrm{B}}\right|^{\mathrm{p}}}, \tag{3.5}$$

where the index indicates the position (s = source and p = pinch magnet). An electron is reflected by the maximal magnetic field, if all its kinetic energy is in form of transversal energy at the pinch magnet $E^{\mathrm{p}} = E_{\perp}^{\mathrm{p}}$ or earlier. Furthermore, with the help of

$$E_{\perp} = E\sin^2(\theta). \tag{3.6}$$

(since $p_{\perp} = p\sin\theta$ and $E \propto p^2$) one finds

$$\frac{E\sin^2(\theta^{\max})}{\left|\vec{\mathrm{B}}\right|^{\mathrm{s}}} = \frac{E^{\mathrm{p}}}{\left|\vec{\mathrm{B}}\right|^{\mathrm{p}}} \tag{3.7}$$

$$\rightarrow \sin(\theta^{\max}) = \sqrt{\frac{\left|\vec{\mathrm{B}}\right|^{\mathrm{s}}}{\left|\vec{\mathrm{B}}\right|^{\mathrm{p}}}} \tag{3.8}$$

For the reference values $\left|\vec{\mathrm{B}}\right|^{\mathrm{s}} = 3.6\,\mathrm{T}$ and $\left|\vec{\mathrm{B}}\right|^{\mathrm{p}} = 6\,\mathrm{T}$ one obtains $\theta^{\max} = 51°$.

Whether an electron can pass the analyzing plane or not does not only depend on its starting energy but also on its starting angle. Accordingly, a β-electron created with a kinetic energy larger than the retarding potential, but under a large angle might thus be reflected. Equation (3.4) shows that some transversal energy still remains in the analyzing plane, which is not analyzed by the retarding potential. Analogously, the maximal transversal energy an electron can still have in the analyzing plane defines the energy resolution ΔE of the spectrometer.

$$\Delta E = E_{\perp}^{a,\max} \tag{3.9}$$

$$= E_{\perp}^{\mathrm{s},\max}\frac{\left|\vec{\mathrm{B}}\right|^{\mathrm{a}}}{\left|\vec{\mathrm{B}}\right|^{\mathrm{s}}} \tag{3.10}$$

$$= E_{\perp}^{\mathrm{p},\max}\frac{\left|\vec{\mathrm{B}}\right|^{\mathrm{a}}}{\left|\vec{\mathrm{B}}\right|^{\mathrm{p}}} \tag{3.11}$$

$$= E\cdot\frac{\left|\vec{\mathrm{B}}\right|^{\mathrm{a}}}{\left|\vec{\mathrm{B}}\right|^{\mathrm{p}}} \tag{3.12}$$

For the parameter settings $E = 18.6\,\mathrm{keV}$, $\left|\vec{\mathrm{B}}\right|^{\mathrm{a}} = 3\cdot10^{-4}\,\mathrm{T}$ and $|\vec{\mathrm{B}}|^{\mathrm{p}} = 6\,\mathrm{T}$ one gets $\Delta E = 0.93\,\mathrm{eV}$. In a MAC-E-filter the term energy resolution has a specific meaning: it implies that an electron starting under the maximal angle $\theta_{\max}$ needs 0.93 eV more kinetic energy than the retarding potential to be transmitted. Accordingly, an electron

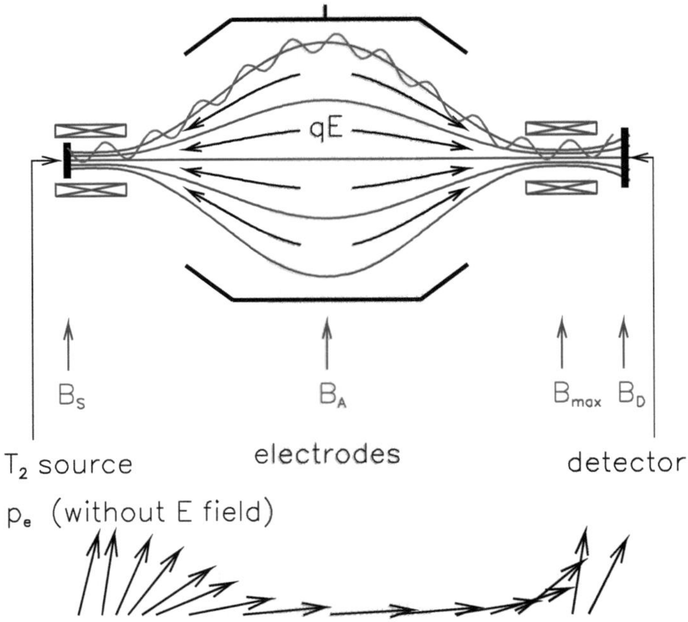

Fig. 3.10 This figure visualizes the MAC-E-Filter principle. A the electrons moves from high (entrance) to low (center) magnetic field its transversal momentum in transformed into longitudinal momentum. The longitudinal momentum is used to overcome the electric retarding potential [3]

emitted parallel to the field lines ($\theta = 0$) only needs an kinetic energy equal to the retarding potential.

3.2.1 The Transmission Function

In the following the transmission probability T as a function of surplus energy (transmission function) will be derived.

The decreasing magnetic field at the main spectrometer entrance transforms transversal energy into longitudinal energy. Simultaniously, the increasing electric potential transforms longitudinal energy into potential energy. An electron can only pass the analyzing plane, if its longitudinal energy at the analyzing plane is larger than zero:

$$E_{\parallel}^{\mathrm{a}} \geq 0 \tag{3.13}$$

This is called the transmission condition. In a first step the maximal angle under which an electron of a certain starting energy must be created in order to be transmitted will be derived. With the help of energy conservation, an expression for the longitudinal energy in the analyzing plane $E_{\parallel}^{\mathrm{a}}$ as a function of the starting energy E^{s} can be found:

$$E^{\mathrm{s}} + qU^{\mathrm{s}} = E^{\mathrm{a}} + qU^{\mathrm{a}} \tag{3.14}$$

$$= E_{\parallel}^{\mathrm{a}} + E_{\perp}^{a} + qU^{\mathrm{a}} \tag{3.15}$$

$$\rightarrow E_{\parallel}^{\mathrm{a}} = E^{\mathrm{s}} - E_{\perp}^{a} + q\left(U^{\mathrm{s}} - U^{\mathrm{a}}\right) \tag{3.16}$$

$E_{\perp}^{\mathrm{a}}$ can be expressed with the help of Eq. (3.4)

$$E_{\parallel}^{\mathrm{a}} = E^{\mathrm{s}} - E_{\perp}^{\mathrm{s}} \frac{\left|\vec{\mathrm{B}}\right|^{\mathrm{a}}}{\left|\vec{\mathrm{B}}\right|^{\mathrm{s}}} + q\left(U^{\mathrm{s}} - U^{\mathrm{a}}\right) \tag{3.17}$$

With Eq. (3.6) one finds

$$E_{\parallel}^{\mathrm{a}} = E^{\mathrm{s}} - E^{\mathrm{s}} \sin^2(\theta) \frac{\left|\vec{\mathrm{B}}\right|^{\mathrm{a}}}{\left|\vec{\mathrm{B}}\right|^{\mathrm{s}}} + q\left(U^{\mathrm{s}} - U^{\mathrm{a}}\right) \tag{3.18}$$

Applying the transmission condition Eq. (3.13) leads to the desired relation between the maximal starting angle and the starting energy:

$$0 = E^{\mathrm{s}} - E^{\mathrm{s}} \sin^2(\theta^{\max}) \frac{\left|\vec{\mathrm{B}}\right|^{\mathrm{a}}}{\left|\vec{\mathrm{B}}\right|^{\mathrm{s}}} + q\left(U^{\mathrm{s}} - U^{\mathrm{a}}\right) \tag{3.19}$$

$$\rightarrow \sin^2(\theta^{\max}) = \frac{E^{\mathrm{s}} + q\left(U^{\mathrm{s}} - U^{\mathrm{a}}\right)}{E^{\mathrm{s}}} \frac{\left|\vec{\mathrm{B}}\right|^{\mathrm{s}}}{\left|\vec{\mathrm{B}}\right|^{\mathrm{a}}}, \tag{3.20}$$

To find the probability for an electron of a certain energy to be transmitted, one has to know the angular distribution of the source. To very good approximation one can assume this to be isotropic. In an isotropic distribution the cosine of the polar angle is equally distributed. Therefore, the fraction T of electron emitted under an angle $\theta < \theta^{\max}$ for an isotropic angular distribution is

$$T(\theta^{\max}) = 1 - \cos(\theta^{\max}) \tag{3.21}$$

From Eq. (3.20) one can find an expression for $\cos(\theta^{\max})$, which can be inserted into Eq. (3.21):

$$T(E^{s}, qU) = 1 - \sqrt{1 - \frac{E^{\mathrm{s}} + q\left(U^{\mathrm{s}} - U^{\mathrm{a}}\right)}{E^{\mathrm{s}}} \frac{\left|\vec{\mathrm{B}}\right|^{\mathrm{s}}}{\left|\vec{\mathrm{B}}\right|^{\mathrm{a}}}} \tag{3.22}$$

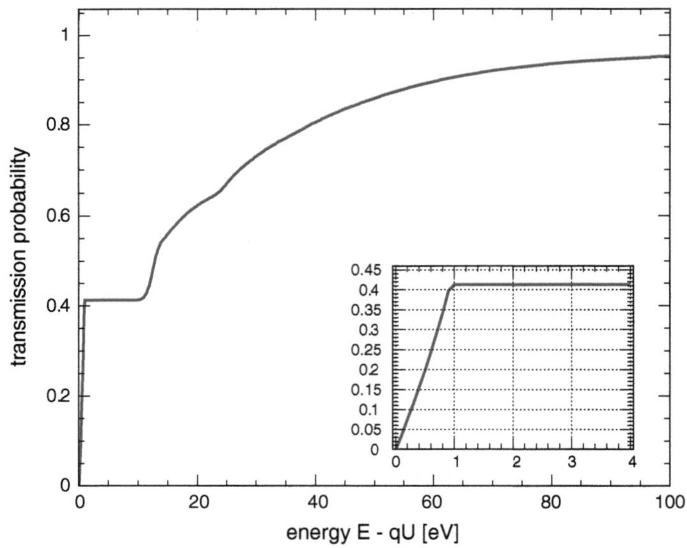

Fig. 3.11 The *curve* shows the transmission probability as a function of surplus energy over the retarding potential. The first part of the function is the transmission function, which is shown on a larger scale in the inset. The bumps at roughly 20 and 30 eV correspond to one-fold and two-fold scattering [3]

Taking into account the area of validity one finally finds

$$T(E^s, qU) = \begin{cases} 0 & \text{for} \quad E^s < q\,(U^s - U^a) \\ 1 - \sqrt{1 - \frac{E^s + q(U^s - U^a)}{E^s}\frac{|\vec{B}|^s}{|\vec{B}|^a}} & \text{for} \quad (U^s - U^a) < E^s < q\,(U^s - U^a) + \Delta E \\ 1 & \text{for} \quad E^s > q\,(U^s - U^a) + \Delta E \end{cases} \tag{3.23}$$

A visualization of the transmission function is given in the inset of Fig. 3.11.

3.2.2 The Response Function

The transmission probability as a function of starting energy for a given retarding potential is given by the transmission function. However, this is only correct if the electron does not loose energy from its point of creation inside the source until it reaches the analyzing plane. However, only about 40 % of all electrons reach the analyzing plane without any inelastic scattering (predominantly in the WGTS). The scattered electrons consequently need a higher starting energy E^s to be transmitted. The response function

$$R(E^s, qU) = \int_0^E T(E^s - \epsilon, qU)\,(P_0\delta(\epsilon) + P_1 f(\epsilon) + P_2(f \otimes f)(\epsilon) + \ldots)\,d\epsilon, \tag{3.24}$$

describing the transmission probability R as a function of starting energy E^s, is a convolution of the spectrometer properties (transmission function T) and the source properties, defining the scattering probabilities P_i and the energy losses $f(\epsilon)$ (see Fig. 3.11). The energy loss distribution $f(\epsilon)$ can be measured in the KATRIN beam line in a specific long-term ($\approx$ 1 month) measurement with mono-energetic electrons from an electron gun. The scattering probabilities depend on the column density ρd of the source which will be monitored by measuring the transmission probability for high surplus energies.

3.3 Systematic and Statistical Uncertainties

Following the principles of a MAC-E filter and the considerations outlined above, KATRIN will measure the convolution of the response function $R(E, qU)$ with the differential energy spectrum $\frac{dN_\beta}{dE}(E_0, m_\nu)$:

$$N(qU, E_0, m_\nu) = N_{\text{tot}} t_U \int_0^{E_0} \frac{dN_\beta}{dE}(E_0, m_\nu) R(E, qU) dE, \tag{3.25}$$

with the parameters N_{tot} and t_U denoting the total number of tritium nuclei and the measurement time at a certain retarding potential. The differential spectrum $\frac{dN_\beta}{dE}(E_0, m_\nu)$ describes the number of decays per second, per nucleus and energy bin. In addition to the signal electrons also background events will be measured. For the fit of the spectral shape, a constant background is assumed. The fit function to the measured spectrum is thus given by

$$N_{\text{th}}(qU, R_s, R_b, E_0, m_\nu) = R_s \cdot N(qU, E_0, m_\nu) + R_b \cdot N_b, \tag{3.26}$$

where R_s and R_b are the relative fraction of signal and background. In the fit R_s, R_b, E_0 and m_ν are free parameters.

3.3.1 Sources of Systematic Errors

A closer look at Eqs. (3.25) and (3.26) reveals potential sources of systematic uncertainties. They can arise from:

- unconsidered corrections to the β-spectrum $\frac{dN_\beta}{dE}(E_0, m_\nu)$
- unaccounted variations of experimental parameters, e.g. retarding potential U and number of decays N_{tot}
- an imprecise knowledge of the response function
- and a non-constant background in time

Systematic effects generally increase if a larger part of the β-spectrum below the endpoint energy E_0 is being used. Furthermore, the effect of the neutrino mass is maximal only in a narrow region about 3 eV below the endpoint, assuming a 10 mHz background rate. This would call for a very narrow measuring interval. However, it is necessary to determine also the endpoint energy E_0 from the measured spectrum. For this purpose, it is statistically advantageous to include a larger part of the spectrum. Therefore, in standard scanning mode KATRIN will measure up to 50 eV below the endpoint energy.

Corrections to the Beta Spectrum

In the β-decay of T_2, both electronic and molecular excitations can occur. Most importantly, the lowest electronic excitation energy in the final state occurs at about 10 eV, so this effect only plays a role for measurement intervals more than 10 eV below the endpoint. If the interval includes only the last 10 eV a constant reduction of the count rate is observed as a consequence of electronic final states. However, since KATRIN is using molecular tritium also rotation and vibration final states occur. The excitation energies of these states fall in the sub-eV regime, thus corresponding to the neutrino mass scale being explored. The final state distribution has been theoretically computed [13, 14]. Systematic effects can arise from uncertainties of this calculation.

Due to the fact that T_2 molecules show a thermal Maxwell-Boltzmann velocity profile as well as a bulk velocity motion in the WGTS, the kinetic energy of the β-electrons is Doppler shifted. With a typical value of the velocity in direction of the emitted electron of $v_{\parallel} = 200\frac{\mathrm{m}}{\mathrm{s}}$, the energy shift is $\Delta E_{\mathrm{doppler}} = 100$ meV. To compute the Doppler shift precisely, the thermal velocity distribution of the molecules, the bulk gas flow and the molecular composition (T_2, DT, etc.) has to be known [15].

If one moves further away from the endpoint E_0 in β-spectroscopy the Fermi function F, taking into account the Coulomb interaction of the outgoing electron with the daughter nucleus, and the nuclear matrix element both become energy dependent. Uncertainties in their theoretical description would lead to a systematic effect of the neutrino mass. However, in the narrow region below the endpoint, where KATRIN is measuring, these effects do not play a significant role.

New physics might of course also affect the shape of the tritium spectrum. However, the influence of these phenomena typically becomes important only further away from the endpoint. New physics effects are for instance the proposed right handed currents [16], sterile neutrinos in the keV range forming warm dark matter, extra dimensions etc. The main new physics contribution that might affect the spectrum also close to the endpoint is violation of Lorentz invariance [17].

Uncertainty of Experimental Parameters

A major experimental challenge arises from the requirement that the retarding potential has to be very stable. The upper limit for the relative allowed variation is $\Delta V/V < 3$ ppm. The variation of the potential at the WGTS has to be limited to $\Delta U < 10$ mV.

Another major challenge is to know the column density ρd of the WGTS very precisely, as the activity of the source and therefore the count rate depends on this parameter. Therefore, it will be measured with an e-gun on a regular basis by measuring the response function at different surplus energies.

In addition, the magnetic field in the source must be stable within $\Delta B_S/B_S < 2 \cdot 10^{-3}$. This is important for two reasons: The transmission function and secondly the response function, i.e. the energy losses, depend on the magnetic field. This stems from the fact that the scattering probabilities depend on the path length of the electrons, which itself depends on the cyclotron radius and hence on the magnetic field.

Imprecise Knowledge of the Response Function

Electrons experience energy losses E_{loss}, primarily due to inelastic scattering (and to a much smaller extent due to elastic scattering) in the WGTS. In addition there are synchrotron radiation losses in high magnetic fields:

- Scattering losses: Energy loss due to elastic scattering is small with $\langle E^{\text{loss, elas}} \rangle = 20$ meV and a cross section which is 10 times smaller than for inelastic scattering. Inelatic scattering leads to energy losses $E^{\text{loss}} > 13.6$ eV.
- Synchrotron losses: Due to their cyclotron motion, electrons loose energy via synchrotron radiation. The maximal synchrotron radiation occurs in the transport section (TS) due to the large magnetic field ($B_{\text{TS}} = 5.6$ T). There the energy loss is $\Delta E_{\text{TS}}^{\text{synch}} = 130$ meV, while the maximal energy loss in the WGTS is $\Delta E_{\text{WGTS}}^{\text{synch}} = 12$ meV.

The energy losses are incorporated in the response function, which is an input to the fit function. Therefore, it has to be measured and simulated very precisely. However, in a region close to the endpoint, the energy losses do not lead to any systematic effect, but only to a reduction of the count rate. In this interval the exact knowledge of the transmission function is essential.

Non-constant Background

First of all, a very low background level is essential to minimize the systematic uncertainties. The higher the background count rate the further away from the endpoint one has to measure the β-spectrum in order to optimize the signal to background ratio. Further away from the endpoint, the impact of systematic effects relative to the influence of a neutrino mass increase.

In the fit function the background is assumed to be constant in both energy and time, therefore, a non constant background level immediately leads to a systematic error.

3.3.2 Sensitivity of KATRIN

For the reference setup of the KATRIN experiment, the quadratic sum of all known systematic uncertainties is expected to be $\sigma_{\text{sys,tot}} \leq 0.017\ \text{eV}^2$. The largest contribution to the error budget from systematics is the description of final states. For determining the statistical errors experimental data are simulated, which take into account all relevant processes. By fitting the theoretical curve to the simulated spectrum, the neutrino mass parameter can be determined. Repeating this procedure many times, one generates a neutrino mass distribution. The width of this distribution then corresponds to the statistical error [15]. The measurement time of KATRIN is chosen such that the statistical error is of the same order as the systematic error. This is reached after three "full beam" years measurement time. Adding systematic and statistical error quadratically yields a total error of $\sigma_{\text{tot}} = 0.025\ \text{eV}^2$.

If no neutrino mass is observed, an upper limit at 90 % confidence level of

$$m(\nu_e) < 200\ \text{meV} \tag{3.27}$$

can be stated. Accordingly, a neutrino mass of $m_\nu = 350$ eV would be seen with 5 σ significance.

3.4 Overview of Background Sources at KATRIN

A summary of all known background sources will be given in this section. Within the frame of this work the main background sources were investigated. With the help of dedicated simulation tools they could be explained, characterized, and countermeasures could be developed. A detailed model of the three main background sources is given in the Chaps. 5, 6, and 7.

The background originates mainly from the spectrometer section and partly from the detector. To achieve the desired sensitivity a background level of $b \leq 10$ mHz is aimed for.

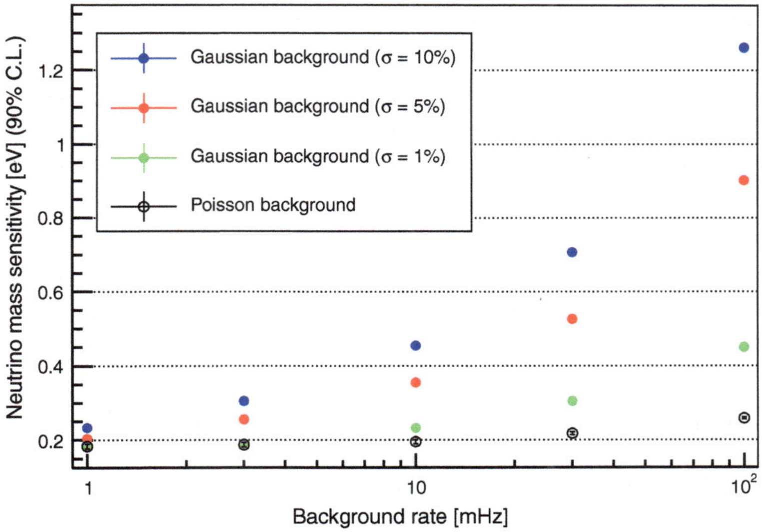

Fig. 3.12 In the best case the background is constant over time and only varies due to natural statistical fluctuations described by a poisson distribution. However, background rates with large fluctuations can be expected from stored electrons. In this plot a simplified Gaussian distributed background model with different width was assumed in order to show the effect of background rates with large fluctuations

3.4.1 Importance of Low Background for KATRIN

Only a small fraction of 10^{-13} of the 10^{11} Tritium decays per second in the WGTS produces β-electrons in the interesting energy region close to the endpoint (1 eV below the endpoint). This leads to the generic low KATRIN count rate of only $10^{-2}\,s^{-1}$ in this energy region. To achieve a sensitivity of 200 meV (90 % C.L.) the background must maximally be of the same order of magnitude.

The 90 % C.L. statistical neutrino mass upper limit depends not only on the absolute background rate but also on the background characteristics. As it was first investigated in this work, especially severe is the effect of a background with non-Poissonian fluctuations. Assuming a Gaussian distributed background with a large width of 5 % and a constant mean of 10 mHz, the neutrino mass sensitivity is decreased almost by a factor of 2 (see Fig. 3.12). As it will be shown in Chap. 7 stored electrons arising from nuclear decays lead to this non-Poissonian distributed background rates.

3.4.2 Detector Background

The detector energy resolution is expected to be of the order of 1 keV. Therefore, all non-signal electrons detected in the region of interest, i.e. between approximately 17 and 19 keV contribute to the background. Possible background sources are electrons produced by

- cosmic muons (and subsequent neutrons and gammas)
- high energetic gammas of environmental radioactivity (mainly thorium and uranium chain in the surrounding area)
- decays of radio-nuclei in the detector material

To reduce the background the construction material is chosen and radio-assayed very thoroughly. In addition, the detector is surrounded by a muon veto system. Finally, there is the option to use post acceleration to move the energy of the signal electrons up to 30 keV, which would allow for better signal background discrimination. The total expected detector background is about 1 mHz, which is verified in the on-going commissioning measurements of the focal plane detector system.

3.4.3 Spectrometer Background

There are several different background production mechanisms in the spectrometer section. In all cases it is important to emphasize that all low energy electrons being created in the volume of the spectrometer or on its inner surface are accelerated on their way to the detector to tank potential and hence, lie in the energy region of interest.

Electrons from the Inner Surface of the Wall and the Wire Electrode

Electrons can be produced on the inner surface of the main spectrometer wall and the wire electrodes mainly via following processes:

- Gammas of several hundred keV from natural radioactivity and high energy cosmic muons penetrate the main spectrometer and subsequently release predominantly low energy electrons from its inner surface.
- Remaining sharp edges of the electrodes or surface roughness of building materials can lead to the process of field emission.
- Radioactive decays of nuclei in the material can produce electrons. The radio-purity requirements of the spectrometer material, however, are very stringent.

Electrons starting on the inner surface of the spectrometer vessel are mostly reflected back to the wall via the magnetic field that is present in the main spectrometer (magnetic shielding). In addition they are repelled by a two layer wire electrode system, that is set to more negative potential than the tank (electric shielding).

In spite of these facts, there is still a chance that surface electrons create background. An electron that is not reflected back to the wall, can be stored in the main spectrometer and, in the presence of non-axially symmetric field components, subsequently drift into the sensitive spectrometer volume. These electrons indirectly create background via ionization.

In Chap. 5 this mechanism will be explained in detail. In particular, two models of the dominating non-axially symmetric magnetic field sources will be presented.

Small Penning Traps

Penning traps are areas in which electrons can locally be stored. A Penning trap is created by a combination of an electric potential well, that traps an electron in axial direction and a magnetic field that confines the electrons in radial direction. Electrons being stored in Penning traps produce background via messenger particles, such as positive ions or photons, which can leave the trap and have the chance to create secondaries in the sensitive volume.

Background rates up to several kHz, caused by a small penning trap with a cross sectional area of only $A \approx 0.75\ \text{cm}^2$, was observed at the pre-spectrometer. A detailed model of the trap filling and background production mechanism, as devised in this thesis, will be described in Chap. 6. As a consequence, the creation of traps must be avoided by implementing an extremely precise electromagnetic design.

Stored Electrons in a Magnetic Bottle Trap

The magnetic field in the main spectrometer drops by four orders of magnitude from the entrance (exit) to the center. This configuration forms a magnetic bottle, causing electrons starting in the volume with a transversal energy of more than $\sim$1 eV to be stored. Owing to the good UHV conditions the storage times of electrons in the multi-keV-range reach several hours. During its cooling time, the stored electron produces several hundred secondary electrons. These secondary electrons are mostly stored again. However, eventually all electrons are cooled down to a level at which they are released from the trap and can hit the detector. In the framework of this thesis the main source of primary electrons was found to originate from ^{219}Rn, ^{220}Rn and Tritium decays inside the spectrometer volume. However, also muon-induced electrons being stored in the main spectrometer volume belong to this category.

A detailed characterization of the ^{219}Rn, ^{220}Rn induced background, based on simulations and measurements at the pre-spectrometer, and the simulation of the expected background from tritium and radon decay in the main spectrometer will be presented in Chap. 7.

Background from β-electrons

Most of the 10^{11} β-electrons created in the source per second are reflected by the pre-spectrometer retarding potential. About 10^4 electrons per second enter the main spectrometer. While residing in the main spectrometer these electrons can ionize residual gas molecules. The resulting positive ions can freely pass the analyzing plane and subsequently produce secondary electrons via ionization. It is those electrons which can hit the detector. This consideration was the prime motive to operate the pre-spectrometer on negative high voltage, since this would reduce the number of electrons entering the main spectrometer.

Trap Between Pre- and Main Spectrometer

The simultaneous operation of two electrostatic retarding spectrometers in a row poses several challenges. Two spectrometers on a high negative potential next to each other, and with ground potential between them, will create a large penning trap for electrons. Signal electrons that experience no energy loss on their way in the spectrometers are not affected. However, any electron that is generated between the spectrometers with less kinetic energy than the pre-spectrometer voltage is trapped. Matching sources can be secondary electrons from ionization, β-electrons that loose energy via scattering, electrons from the wall, etc.

A stored electron will fill the trap in an avalanche effect: by continually ionizing residual gas molecules, it produces more electrons which are also stored in the trap, thus producing more secondaries and so on. The cooling time of a 18 keV stored electron is about 1 h at a pressure of $p = 10^{-11}$ mbar. A primary stored electron together with all its secondaries, will create up to 100 Million positive ions. As positively charged particles are not trapped, they can freely propagate into the main spectrometer. There, they will ionize residual hydrogen molecules with a probability of about 10^{-7}, thereby producing further electrons that can reach the detector.

Assuming a filling rate of the trap with primaries of 1 kHz, one can expect a resulting background rate of at least 10 kHz at the detector [18]. This exceeds the desired limit of 10 mHz by six orders of magnitude, so highly effective countermeasures have to be adopted.

One option would be a wire scanner, installed between the spectrometers, that regularly wipes through the trap, thereby removing the stored electrons and suppressing the background production [19]. Another more drastic, yet most promising option would be to operate the pre-spectrometer at zero potential, which would avoid the creation of the trap in the first place. Simulations demonstrate that the background expected from signal electrons penetrating the main spectrometer is much smaller than the background expected from the Penning trap, making this proposal a very attractive solution to a severe background problem.

3.4.4 Countermeasures for Background

For most of the background processes mentioned above there is a direct countermeasure to be applied:

- The magnetic and electric shielding offers an almost complete reduction of the muon- and gamma-induced background from the inner vessel walls.
- Background from small Penning traps can be avoided by carefully designing all components so that no Penning traps are created ab initio.
- Background from signal electrons is a rather negligible contribution. It can be reduced by setting the pre-spectrometer on high potential.
- Setting the retarding potential of the pre-spectrometer to zero or the usage of a wire scanner eliminates the background from the penning trap between the spectrometers.

However, there remains one major source of background, that is not tackled with the countermeasures mentioned above. Stored electrons from nuclear decays in the spectrometer (or to lesser extent, from cosmic muons) can generate substantial levels of background. Corresponding simulations based on test measurements with the pre-spectrometer revealed that the expected background exceeds the desired limits by at least one order of magnitude.

As a consequence, the allowed number of tritium decays in the spectrometer has to be drastically reduced compared to the former design values. Furthermore, single ^{219}Rn atoms, mainly emanating from the NEG-getter pumps, need to be be passively shielded from the main spectrometer by a nitrogen cooled baffle. However, the isotope ^{220}Rn emanating from structural material used in the interior of the main spectrometer cannot be reduced by these means.

Combining all passive reduction methods, the expected rates still exceed the limit of 10 mHz, therefore further active reduction methods have to be implemented. The goal of these methods will be to remove stored electrons over a wide energy range, without either increasing other background sources nor interfering with the neutrino mass measurement. The most promising method for this task, developed in the framework of this thesis, is based on stochastic heating of stored electrons by the method of Electron Cyclotron Resonance (ECR), as described in Chap. 8.

References

1. "MARE homepage". http://mare.dfm.uninsubria.it/frontend/exec.php [last update: 18-Mar-2011]
2. B. Monreal, J.A. Formaggio, Relativistic cyclotron radiation detection of tritium decay electrons as a new technique for measuring the neutrino mass. Phys. Rev. D **80**, 051301 (Sep 2009)
3. K. Collaboration, KATRIN Design Report (FZKA Report 7090), Tech. Rep., KIT, 2004
4. S. Grohmann, T. Bode, H. Schön, M. Süsser, Precise temperature measurement at 30 K in the KATRIN source cryostat. Cryogenics **51**(8), 438–445 (2011)

5. G. Drexlin, Collection of figures from talks. Private communication
6. A. Windberger, *Berechnungen und Simulationen zum Verhalten von Ionen in der differenziellen Pumpstrecke des KATRIN-Experiments* (Master's thesis, KIT, 2010)
7. M. Ubieto-Díaz, D. Rodríguez, S. Lukic, S. Nagy, S. Stahl, K. Blaum, A broad-band FT-ICR Penning trap system for KATRIN. Int. J. Mass Spectrom **288**(1–3), 1–5 (2009)
8. F. Glück, "Electrons and ions in KATRIN STS". KATRIN Collaboration meeting, October 2010
9. F. Fränkle, Background Investigations of the KATRIN Pre-Spectrometer. PhD thesis, KIT, 2010
10. "KATRIN homepage". http://www-ik.fzk.de/tritium/. Last update 26 Aug 2010
11. T. Thümmler, Präzisionsüberwachung und Kalibration der Hochspannung für das KATRIN-Experiment.PhD thesis, Universität Münster, 2007
12. P. Renschler, KESS - A new Monte Carlo simulation code for low-energy electron interactions in silicon detectors. PhD thesis, KIT, 2011
13. N. Doss et al., Molecular effects in investigations of tritium molecule β-decay endpoint experiments. Phys. Rev. C **73**(2006), 025502 (2006)
14. N. Doss, J. Tennyson, Excitations to the electronic continuum of $3HeT^+$ in investigations of the T_2 beta-decay experiments. J. Phys. B: At. Mol. Opt. Phys. **41**(2008), 125701 (2008)
15. M. Hötzel, *Berechnung von KATRIN Messspektren unter Einbeziehung der fensterlosen gasförmigen Tritiumquelle* (Master's thesis, KIT, 2009)
16. J. Bonn, K. Eitel, F. Glück, D. Sevilla-Sanchez, N. Titov, K. Blaum, The KATRIN sensitivity to the neutrino mass and to right-handed currents in β-decay. Phys. Lett. B **703**(3), 310–312 (2011)
17. J. Carmona, J. Corts, Testing Lorentz invariance violations in the tritium β-decay anomaly. Phys. Lett. B **494**(1–2), 75–80 (2000)
18. F. Glück, "Background at FPD and the pre-spectrometer potential". http://fuzzy.fzk.de/bscw/bscw.cgi/d477645/95-TRP-4518-B1-FGlueck.ppt
19. M. Beck, K. Valerius, J. Bonn, K. Essig, F. Glück, H. Ortjohann, B. Ostrick, E. Otten, T. Thümmler, M. Zbořil, C. Weinheimer, Effect of a sweeping conductive wire on electrons stored in a Penning-like trap between the KATRIN spectrometers. Eur. Phys. J. - Hadrons and Nuclei **44**, 499–511 (2010). doi:10.1140/epja/i2010-10959-1

Chapter 4
Monte Carlo Simulation Package

Given the complexity of the background processes described above, a highly versatile, fast and precise simulation code is required. KASSIOPEIA is the primary simulation package for the KATRIN experiment fulfilling all these requirements. It is written in C++, and comprised of specific modules for the creation, trajectory calculation in electro-magnetic fields and detection of particles in Si-based detectors. A variety of different particle generators (Sect. 4.3.1), including a detailed WGTS model (Sect. 4.3.5) are included. Different electric and magnetic field calculation (Sect. 4.3.3) methods, as well as different tracking modes (Sect. 4.3.2) are available. Physical processes like synchrotron radiation and scattering can be taken into account. The particle detection module includes backscattering of electrons on the detector surface as well as a comprehensive number of physical phenomena of low energy-electrons in silicon (Sect. 4.3.4).

In the framework of this thesis the main focus was on developing the system architecture, user interface, and data output of KASSIOPEIA in collaboration with co-workers from the Massachusetts Institute of Technology (MIT). Besides the general framework, special emphasis of this work was on the development of KTRACK, the particle trajectory calculation package. Moreover, this work was strongly involving the development of several particle generators, such as radon and krypton event generators.

In this chapter the versatile applications of KASSIOPEIA will be outlined. In Sect. 4.2 the general organization of the code will be described and in Sect. 4.3 the individual modules of KASSIOPEIA are explained in detail. Finally, an insight on the user interface and the output is given in Sects. 4.4 and 4.5.

4.1 Purpose of Kassiopeia

The Monte Carlo simulations performed with KASSIOPEIA are applicable for several purposes, described in the following.

S. Mertens, *Background Processes in the Electrostatic Spectrometers of the KATRIN Experiment*, Springer Theses, DOI: 10.1007/978-3-319-01177-6_4,

4.1.1 Optimization of Electromagnetic Design

KASSIOPEIA is an essential tool for the optimization of the electromagnetic design of KATRIN. With the help of the both fast and precise field calculation methods [1, 2], the electric potential as well as the electric and magnetic field in the entire KATRIN setup can be calculated. KASSIOPEIA provides a large variety of different field calculation methods, ranging from very fast axisymmetric field calculations, to fully three-dimensional field calculation methods (for more detail in 4.3.3). Specific issues related to the electromagnetic design optimization can be found in Appendix A.

4.1.2 Monte Carlo Simulations

Besides precise and fast field calculation methods, KASSIOPEIA provides advanced algorithms to compute particle trajectories in electro-magnetic fields down to the level of machine precision.

This tool of KASSIOPEIA allows to perform Monte Carlo simulations of specific measurements. In particular during the design and commissioning phase of KATRIN the experimenter has the ability to make full use of KASSIOPEIA to better understand the results of test experiments. These test measurements at present mostly address background processes, and transmission properties of the spectrometer. How KASSIOPEIA was used for modeling of background mechanisms will be presented in detail in Chaps. 5, 6, 7 and 8. An example of a detailed simulation of a transmission function measurement is described in Appendix B.

4.1.3 Investigation of Systematic Effects and Statistical Uncertainty of KATRIN

KASSIOPEIA provides a detailed tritium source model, that allows to simulate the actual neutrino mass measurement with high precision. The source model includes, among other things, the final state distribution of tritium, the Doppler broadening and scattering in the source (more details in Sect. 4.3.5). Furthermore, KASSIOPEIA includes classes for fitting the integrated tritium β-spectrum and thereby for determining systematic effects and the neutrino mass sensitivity.

With this functionality of KASSIOPEIA it is possible to investigate systematic effects, for instance arising from shifts of the magnetic field or electric potential, the imprecise knowledge of the final state distribution or unstable column density, etc. Furthermore, the statistical uncertainty of KATRIN can be studied with KASSIOPEIA. The investigation with KASSIOPEIA of the effect of strongly fluctuating background on the KATRIN sensitivity will be presented in Chap. 8.

4.2 General Organization

In the following an overview of the basic structure and generic organization of KASSIOPEIA is given. The description is divided into two parts: First the organization of the information produced during the simulation is explained. The second part deals with the question of how this information is produced. The interplay of the different parts of KASSIOPEIA is visualized in Fig. 4.1.

4.2.1 *Representation of Physical States*

KASSIOPEIA saves the information produced during the simulation in specific data containers, which are organized into four physically intuitive levels of detail: Runs, Events, Tracks, and Steps.

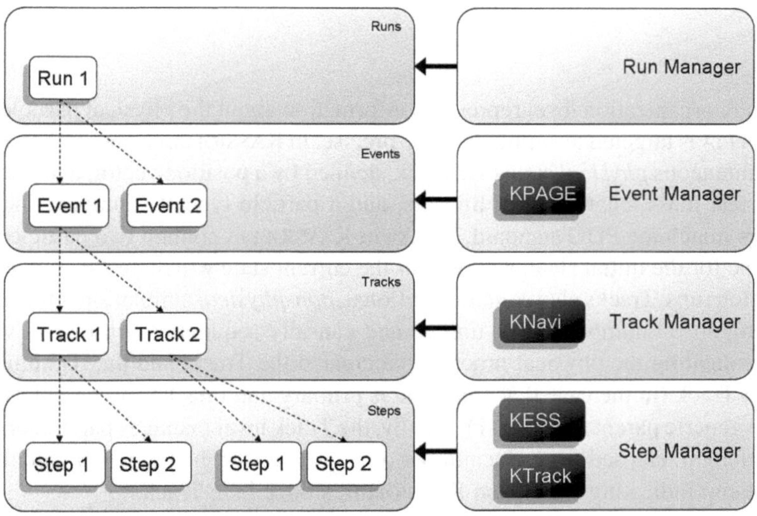

Fig. 4.1 General structure of Kassiopeia. The information produced in a KASSIOPEIA simulation is stored in so-called data containers organized in a different levels of detail: Run, Event, Track, Step. The physical modules, such as KPAGE for particle generation, KNavi for navigating the particle, and KTrack and KESS for the actual step computation, produce the information stored in the corresponding data containers. Managers on all levels of detail organize the execution of the physical modules, see text

Runs

The Run organization level ranks the highest level, and gathers everything that happens during a single execution of KASSIOPEIA. A Run pools an ensemble of Events and stores global information that only pertains to Runs. This includes the number of Events that occur in a Run, and the amount of physical time elapsed during the period of the Run. It is important to note that no physical parameters of the simulation may change during a Run.

Events

The primary particles created by an event generator and all their secondary particles represented by Tracks are grouped together into an Event. The Event also contains general information such as a creation ID, the number of Tracks, and initial and final time of the Event.

Tracks

The Track organization level represents information about the physical particles that KASSIOPEIA is targeted to simulate. To be precise, in KASSIOPEIA a particle represents an instantaneous *physical* state of a Track, defined by a position vector, a momentum vector, rest mass, charge, spin, lifetime, and a particle type ID number. These ID numbers match the PDG standard. Tracks in KASSIOPEIA contain two particle states each: one for the initial state, and one for the current state which gets updated as the simulation runs. Tracks also contain additional, *non-physical* simulation information, including an ID number that is unique and causally sequential within an Event, a stamp indicating the physical process that created the Track, and the ID number of a parent Track (in the case that a particle is primary and thus has no parent Track it gets the generic parent track ID -1). Finally, the Track level manages parameters such as path length, elapsed time, the number of steps used for the trajectory calculation, and a stamp indicating the reason KASSIOPEIA stopped the Track.

Steps

The Step organization level is the finest level of detail considered in KASSIOPEIA. A Step is an incremental, discrete change to a Track. In most applications it has two conceptual parts, the first being the numerical solution to an equation of motion, and the second being the simulation of discrete physical processes that may have occurred during flight along the step. Steps keep track of their initial physical state, again represented as a particle. Their final physical state, which, if the Step is acceptable, becomes the most current physical state of the Track to which the Step belongs.

4.2.2 Evolution of Physical States

Two types of classes are needed to evolve the state of the simulation: On the one hand there are Modules, which initialize and update the contents of the data containers, on the other hand there are Managers that organize and control which Modules are active at a specific time.

Modules

To fill the data containers so-called Modules are required, which have the task to

- create events
- compute the trajectory of each created particle
- localize the particle with respect to the geometric setup of the simulation.

Devoted to the creation of the events are the so-called Generator Modules, comprised in the package KPAGE. For calculating the trajectory of a particle several Step Strategy Modules are necessary. Three types can be distinguished:

- Process type of Module, which is responsible for actually calculating the final particle states within a Step.
- Step Size type of Module, which cooperate to determine the largest possible time increment over which a set of Processes may accurately act.
- Exit Condition type of module, which can indicate that a Step is the final one KASSIOPEIA will calculate for a given Track.

The suite of Processes, Step Sizes and Exit Conditions are comprised in the package KTRACK. Another way to compute trajectories is with KESS, which is a specifically designed process used for tracking in silicon. Other Modules necessary for trajectory calculations are Field Modules, which define the present electric and magnetic fields and the method to compute them. All Modules needed at any point during the simulation are stored in the so-called Toolbox classes.

KASSIOPEIA has the ability to change the trajectory and field computation methods, depending on the position of the particle. The user can define so-called Regions, in which Step Strategy and Field Modules can be switched on or off. Thus every region has a list of entry and exit commands, which are executed once the particle enters (or leaves) a region. The opposite command is executed when the particle leaves (or enters) the region, respectively. A set of Regions in KASSIOPEIA is always a properly nested set of volumes, meaning that any Region may have sub-Regions which are completely contained within the parent Region without overlapping its boundaries.

As a Track is being computed, KASSIOPEIA needs to know which Region the Track's current Particle state falls inside in order to load the proper Step and Field Modules. This is accomplished using a piece of code called the Navigator, which acts like an iterator for the tree structure of the regions.

All Modules that are currently active are stored in the so-called Root Classes. Hence, the content of the Root Classes will be changed while the program is running, whereas the content of the Toolboxes is fixed at the initialization phase of the program.

Managers

The main managers are the Run, Event, Track, and Step Manager. The Run Manager specifies how many events will be created and calls the Event Manager successively to create this number of events. The Event Manager in turn tells the Event Generator Modules to create events, i.e. a number of primaries and passes them onto the Track Manager, where first of all the Navigator is called to orientate the particle. As long as no Exit Condition is met the particles are passed on to the Step Manager where the active Step Strategy Modules are executed. After each step the Track Manager again calls the Navigator to check whether the particle has entered or exited a specific Region.

4.3 Overview of Physical Modules

In this section the modules of KASSIOPEIA being responsible for the creation, the tracking and the detection of particles will be presented. Furthermore, vital tools like electro-magnetic field solving methods and the module for the source spectrum calculation will be described.

4.3.1 Particle Creation

The module subserving the particle generation is the KAssiopeia PArticle GEnerator: KPAGE [3]. The generic parameters determining the starting conditions of a particle comprise its starting time, position, energy and direction. Correspondingly, the structure of KPAGE is based on a modular assembly system. It provides separate time, position, energy and direction creators. These creators can arbitrarily be put together like building blocks to form a full particle generator. Correlated starting parameters are taken into account by appropriately pre-combined creators.

The user can adjust KPAGE to create particles at a fixed position, homogeneously distributed on a surface, in a volume, or adjusted according to the tritium source dynamics. Particles can be created at a fixed time, with a constant rate or with an exponential decay time distribution. The particle's starting energy may be mono-energetic, equally or Gaussian distributed within a user defined interval, or adjusted to the krypton, radon or tritium decay energy spectrum. Finally, the starting direction of particles can be chosen to be fixed, isotropic, or following an isotropic emission from a surface, Gaussian distributed or corresponding to a measured angular distribution (e.g. emission from gold surface, for e-gun simulations).

Example: Radon event generator

The Radon Event Generator belongs to the group of energy creators. As explained in detail in Sect. 7.2.2 the radon isotopes ^{219}Rn and ^{220}Rn undergo α-decays into excited polonium states, which de-excitates instantaneously. Within the decay sequence, several processes can lead to electron production. The Radon Event Generator emulates the following physical mechanisms:

- Shake-off electrons
- Conversion electrons
- Auger electrons
- Shell reorganization electrons.

While the physics model is discussed later, here the focus is on the implementation of the different electron creation mechanisms into the Radon Event Generator.

Based on data in [4, 5] a random generator is used to decide whether a conversion electron is created. If so, an electron from the conversion electron energy spectrum is created. Tables D.1 and D.2 in the Appendix D show the conversion electron energies and conversion electron creation probabilities for ^{219}Rn and ^{220}Rn, which serve as input here.

Similarly, a random generator is used to decide whether a shake-off electron is created, in this case the data is take from [6, 7], as shown in Table D.4. According to [8] the shake-off energy distribution follows a power law

$$N(E_{\text{shake-off}}) = \left(\frac{E_{\text{b}}}{E_{\text{b}} + E_{\text{shake-off}}} \right)^8, \tag{4.1}$$

where E_{b} denotes the binding energy of the corresponding electron shell.

Shake-off and conversion electrons leave vacancies in the inner electron shells. When filling these vacancies with electrons from higher shells, Auger and Coster-Krönig electrons can be created, which themselves leave vacancies. This cascading procedure is simulated based on data taken from PENetration and Energy LOss of Positrons and Electrons (PENELOPE) [9, 10].

If neither conversion nor shake-off electrons were produced, the atom would remain in a double negatively charged final state. Measurements [11] show that the total final charge of the recoiling polonium is most likely zero. As a consequence, two shell reorganization electrons are emitted from the outer shell, which share the energy of $E_{\text{s.r.}} = 230$ eV, based on calculations in [12].

The Radon Event Generator is applicable for the isotopes ^{219}Rn and ^{220}Rn. The corresponding data tables are loaded respectively. Figure 4.2 shows the electron energy distributions of ^{219}Rn and ^{220}Rn α-decay, simulated with the Radon Event Generator.

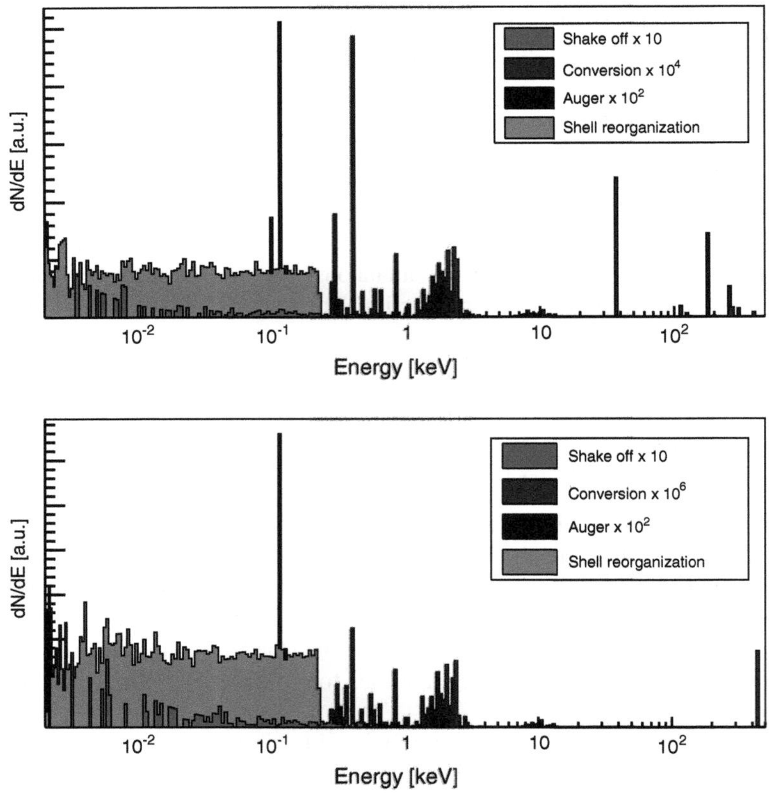

Fig. 4.2 Electron energy spectrum of ^{219}Rn and ^{220}Rn α-decay. The color codings indicate the different mechanisms resulting in the emission of electrons from the eV to the multi-keV scale. *Top* Energy spectrum of ^{219}Rn α-decay. *Bottom* Energy spectrum of ^{220}Rn α-decay. The generation of high-energy conversion electrons in the case of ^{220}Rn α-decay can be almost neglected

In Sect. 7.2.3 the described radon model will be validated by comparing the simulated data of the Radon Event Generator to pre-spectrometer data, as well as to data from the literature. This example emphasizes the approach of this thesis, which is to build a complete physics model of the background process being investigated.

4.3.2 Particle Tracking (KTrack)

KTRACK is a program package to compute particle trajectories. The program is valid for any type of particle.

KTRACK provides two main computation modes: The first tracking mode is an exact trajectory calculation based on solving the Lorentz equation. The second tracking mode is based on computing the guiding center motion of the particle, making use of the adiabaticity of the motion, i.e. the conservation the orbital magnetic moment. Here, the magnetron drift motion and gyration can be taken into account in terms of an analytic approximation. The adiabatic method has the potential to be much faster compared to the exact method, since it allows for much larger step sizes.

In both stepping modes processes such as synchrotron radiation, elastic and inelastic scattering on hydrogen, nitrogen, water and argon can be taken into account.

To solve the equation of motion the user can choose, among several different ordinary differential equation (ODE) solving methods, some of which provide an intrinsic time step control. KTRACK additionally provides a number of criteria for determining the time step, e.g. by means of energy conservation, a fixed step length, limits on the energy loss due to synchrotron radiation, and limits on scattering probability. Any combination of controls can be chosen.

Finally, KTRACK provides a large selection of criteria to terminate a trajectory, for instance by defining a maximal number of steps, maximal path length, surface hit, etc.

To allow for fast calculation KTRACK includes a rather sophisticated caching system, that assures that no quantity is computed twice if not necessary. This is especially necessary for the computationally expensive electromagnetic field calculations.

In the following the most important pieces of KTRACK will be described in detail.

Processes

KTRACK is organized in a general modular framework. The basic unit of this framework is called a Process, which generally represents any action that changes the state of a particle. For example, the propagation of a particle according to an equation of motion is a kind of Process, called Propagation. It changes the particle's position and momentum. The effect of synchrotron radiation loss during a step is likewise encapsulated in a Process, called Synchrotron. Its action is to reduce the transverse momentum of a particle.

Processes in KTRACK are combined into composite tree structures. The exact and adiabatic stepping method, represented by the Exact Step Computer and Adiabatic Step Computer classes are themselves treated as a Process in KTRACK. They sit at the base of a tree whose branches represent sub-Processes (see Fig. 4.3). Possible branches are

- Propagation,
- Synchrotron,
- Drift (only for adiabatic step computer),
- Gyration (only for adiabatic step computer), and
- Scattering.

All processes can in principle have their own sub-subprocesses. This tree-like structure is realized in C++ according to the Composite pattern.

In the current version of KTRACK continuous processes, such as synchrotron radiation, magnetron drift and gyration motion are treated as first order corrections

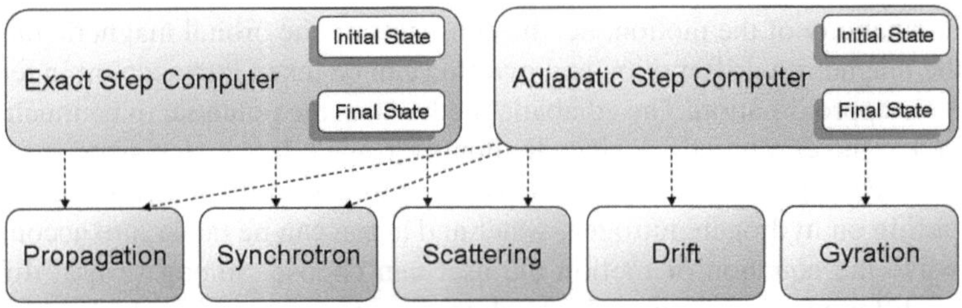

Fig. 4.3 Processes in KTrack. Both step computers, themselves being processes, hold a vector of physical processes and a pointer to the state of the particle at the beginning of the step and during the step

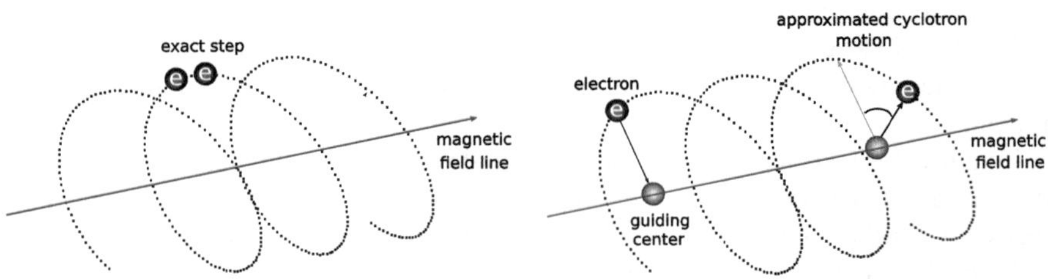

Fig. 4.4 Sketch of two different stepping modes. *Left* The exact step is computed by numerically solving the Lorentz equation (4.2), (4.3). *Right* The guiding center motion is calculated by numerically solving Eqs. (4.4) and (4.5). Drift motion and gyration are added analytically

to the propagation. Likewise, discrete processes like scattering are executed after the propagation process has been completed. This allows for fast trajectory calculations and is of sufficient accuracy. In the up-coming version of KTRACK, however, the user will have the option to integrate continuous processes into the ODE and to locate the actual point of scattering during a step.

Step Computers Two different stepping modes, represented by the Exact Step Computer and Adiabatic Step Computer classes, are available. In the exact stepping mode the particle trajectory is computed by numerically solving the Lorentz equation. With the Lorentz equation the motion of a charged particle in electromagnetic fields is fully described (i.e. cyclotron motion, drift motion, and non-adiabatic motion are included). In the adiabatic stepping mode the particle trajectory is computed based on its guiding center motion. At the beginning of the track the particle's guiding center is determined, and its motion is computed by solving the "Guiding Center ODE". Drift motions are added after the execution of the guiding center in an approximative way. After each step the real particle position is computed by taking into account the cyclotron motion, likewise in an approximative way (see Fig. 4.4). Both stepping modes allow for adding synchrotron radiation and scattering.

Propagation The major Process is the propagation of the particle. This Process involves solving the equation of motion. The ODE in case of the exact stepping mode is the Lorentz equation

$$\dot{\vec{x}} = \vec{v} \tag{4.2}$$

$$\dot{\vec{v}} = q\left(\vec{E} + \vec{v} \times \vec{B}\right) \tag{4.3}$$

In case of the adiabatic approximation an ODE describing the guiding center motion of the particle is used

$$\dot{\vec{x}}_{\mathrm{GC}} = \hat{B} \cdot v_{\mathrm{L}} \tag{4.4}$$

$$\dot{v}_{\mathrm{L}} = -\frac{\mu}{\gamma}\left(\vec{\nabla}|\vec{B}|\right)\hat{B} + q\vec{E} \cdot \hat{B} \tag{4.5}$$

In case of the exact stepping mode the particle's position $\vec{x}$ and velocity $\vec{v}$ are altered with time whereas the guiding center ODE describes the time evolution of the guiding center position $\vec{x}_{\mathrm{GC}}$ and the longitudinal velocity v_{L}. Three types of ODE solvers are available:

- *Runge Kutta*: KTRACK has implemented a 8th order Runge-Kutta integrator, requiring 13 evaluations per step [13, 14].
- *Embedded Runge Kutta*: This method combines two different Runge-Kutta integrators of different order allowing for an internal error estimation and step size control, accordingly. Currently implemented are (4th/5th *order Runge-Kutta*, 5th/6th *order Runge-Kutta*, 6th/8th *order* and 8th/7th *order Runge-Kutta*) [15, 16].
- *Predictor Corrector*: This is an Adams-Bashforth predictor utilizing a modified Adams-Moulton corrector requiring two field evaluations per step [17]. This solver is capable of estimating numerical errors.

Synchrotron The synchrotron process computes the average energy loss $\langle E_{\mathrm{loss}} \rangle$ due to synchrotron radiation from the particle properties at the beginning and end of the step. With $\langle E_{\mathrm{loss}} \rangle$ the transversal momentum of the particle is updated.
Scattering Scattering is composed in a composite tree structure. The basic scattering class holds a vector of scattering submodules, that simulate scattering on different molecules. For each scattering submodule elastic and inelastic scattering can be individually added. When executing the scattering process, it is first decided whether scattering takes place at all. If so, it is decided on which module (i.e. scattering on hydrogen or any other activated molecule). Then the scattering off the active module is executed, whereby it is decided which kind of scattering actually takes place (i.e. elastic, ionization, excitation). Once decided, the specific submodule is finally executed to compute the energy loss and the angle change of the primary particle and possibly to create a secondary particle.
Drift and gyration In case of the adiabatic stepping mode, two more processes can be added: Drift motion and gyration. To calculate the drift motion first the average drift velocity $\langle \vec{v}_{\mathrm{drift}} \rangle$ is computed, then the guiding center position of the particle is updated accordingly. To take the gyration of the particle into account the azimuthal phase change $\Delta\phi$ is computed. With $\Delta\phi$ the vector pointing from the guiding center to the real particle position is computed and the particle position is accordingly updated.

Step size controls

As mentioned above, there are various criteria to control the step size available in KTRACK. The basic behavior of these controls is captured in the Step Size classes. Each Step Size class naturally belongs to a particular Process class. For instance, the energy conservation step control naturally belongs to Propagation, and the synchrotron radiation step control naturally belongs to Synchrotron. Each of these controls are able to suggest a time step, which is that control's best guess of the maximum time step it itself can tolerate. Each control is also able to examine whether the propagation with the suggested time step has exceeded the control's tolerance, at which point it can veto the step and make an improved suggestion of the time step.

The following step size controls are available:

- propagation associated step size control: defining the maximal
  - numerical error
  - fraction of cyclotron period
  - energy conservation violation
  - allowed step length
  - allowed time step
- synchrotron associated step size control: defining the maximal allowed energy loss due to synchrotron radiation
- scattering associated step size control: defining the maximal allowed scattering probability.

All these step size controls can be independently used and adjusted by the user.

In the beginning of a step all active step size controllers are asked to suggest a time step. Then, the propagation will be performed with the smallest of all suggestions. At the end of executing the propagation, the step size controls are asked to check whether this step exceeds any of the given limits. (e.g. if the violation of energy conservation was really within the allowed range). If one of the checks fails, the step is repeated with a smaller time step, which is suggested from whichever step controller failed until all checks are successful. Then processes like Synchrotron and Scattering can be executed.

Exit Condition

After each step it is checked whether an exit condition is met. The available exit conditions are

- maximal number of steps,
- maximal time of flight,
- maximal path length,
- minimal energy,
- maximal number of axial turns (this applies to trapped particles),

- maximal number of magnetron turns (this applies to trapped particles),
- minimal distance to a surface,
- particles enters/leaves specific Region.

4.3.3 Field Calculation Methods

KASSIOPEIA provides a comprehensive number of field calculation methods. In the following the basic principles of the methods will be presented.

Magnetic field calculation

The main sources of magnetic fields at KATRIN are normal conducting and super conducting coils, all of which are axially symmetric around their axis. However, both magnetic material, and the earth magnetic field compensation system (EMCS) are non-axially symmetric field components that need to be taken into account, too.

Axially symmetric magnetic field calculation For the axially symmetric magnetic field calculation two methods are available. The elliptic integral method, which is valid over a large spatial range, however, needs rather long calculation times. The computation time can be reduced by a factor of up to hundred by switching to a second method, the Legendre polynomial expansion method. At KATRIN, the electrons usually move inside the magnetic flux tube, which keeps an adequate distance to the coils. Therefore, in almost all cases the fast Legendre polynomial computation method is appropriate.

Elliptical integrals Axially symmetric coils of a certain dimension can be build up by circular infinitesimally thin current coils. The magnetic field of a thin current loop can be expressed in terms of elliptic integrals [18]

$$K(k) = \int_0^{\frac{\pi}{2}} \frac{\mathrm{d}\varphi}{\sqrt{1 - k^2 \sin^2 \varphi}} \tag{4.6}$$

$$E(k) = \int_0^{\frac{\pi}{2}} \mathrm{d}\varphi \sqrt{1 - k^2 \sin^2 \varphi} \tag{4.7}$$

$$\Pi(c, k) = \int_0^{\frac{\pi}{2}} \frac{\mathrm{d}\varphi}{(1 - c^2 \sin^2 \varphi)\sqrt{1 - k^2 \sin^2 \varphi}} \tag{4.8}$$

as

$$B_r = \frac{I}{c} \cdot \frac{2z}{r\sqrt{(a+r)^2+z^2}} \left[-K(k) + \frac{a^2+r^2+z^2}{(a+r)^2+z^2} E(k) \right] \tag{4.9}$$

$$B_\varphi = 0 \tag{4.10}$$

$$B_z = \frac{I}{c} \cdot \frac{2}{r\sqrt{(a+r)^2+z^2}} \left[K(k) + \frac{a^2-r^2-z^2}{(a+r)^2+z^2} E(k) \right]. \tag{4.11}$$

where ϕ denotes the azimuthal angle, r the radius and z the axial position of the field point. The parameter a denotes the radius of the coil and I denotes its current. The parameters c and k are functions of a, r, and z. To compute the magnetic field of a coil with real dimensions two more steps have to be performed. To take into account the length L an analytic expression can be found by using the third elliptic integral $\Pi(c, k)$. Finally, a numerical integration over the radius of the coil (from inner radius to the outer radius) has to be performed.

Legendre polynomial expansion A much faster method is based on the Legendre polynomial expansion, also known as zonal harmonic expansion [2]. Here the field at any point in the area of validation (which will be explained below) can be computed as an expansion in Legendre polynomials. The necessary coefficients, the so-called source coefficients, are prepared beforehand and are valid as long as the coil setup does not change. The same source coefficients can be used for all field points within the area of validation, taking into account a radius correction. The magnetic field in terms of Legendre polynomials is given by

$$B_r = -s \sum_{n=1}^{\infty} \frac{B_n^{\text{cen}}}{n+1} \left(\frac{\rho}{\rho_{\text{cen}}} \right)^n P_n'(u) \tag{4.12}$$

$$B_\varphi = 0 \tag{4.13}$$

$$B_z = \sum_{n=0}^{\infty} B_n^{\text{cen}} \left(\frac{\rho}{\rho_{\text{cen}}} \right)^n P_n(u) \tag{4.14}$$

$$\text{with} \quad u = \cos\theta \quad \text{and} \quad s = \sin\theta \tag{4.15}$$

with B_n^{cen} being the central source coefficients and P_n the Legendre polynomials. The smallest distance between the source point and the coil ρ_{cen} is usually called central convergence radius and Eq. (4.12) is only valid within (see Fig. 4.5).

To maximize the area of validation, many source points are used. Moreover, the smaller the ratio of the parameters $\frac{\rho}{\rho_{\text{cen}}}$ the faster the expansion converges. With more source points the chance is higher to find a small ratio for every field point. Hence, the computation time can be reduced.

To be able to compute the magnetic field outside of the coil the remote Legendre polynomial expansion is used. The remote convergence radius ρ_{rem} is given by the maximal distance of the source point to the coil, and the field calculation is valid for all points with $\rho > \rho_{\text{rem}}$. The areas of validation are visualized in Fig. 4.5.

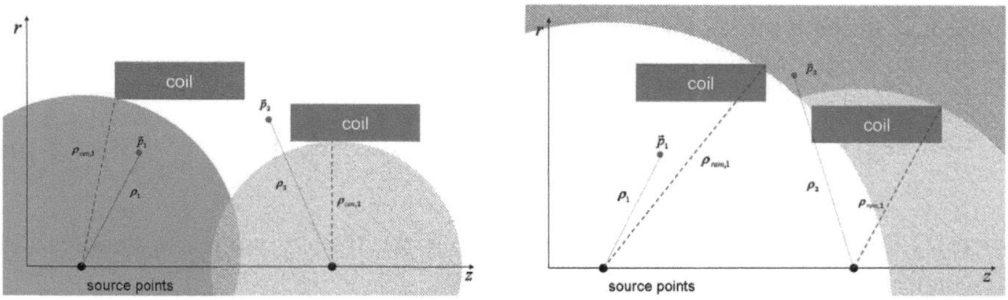

Fig. 4.5 Area of validation for Legendre polynomial expansion. *Left* At the field point $\vec{P}_1$ the *central* Legendre polynomial expansion is valid, whereas $\vec{P}_2$ lies outside of the range of validity. *Right* At the field point $\vec{P}_2$ the *remote* Legendre polynomial expansion is valid, in this case $\vec{P}_1$ lies outside of the range of validity

For tilted coils, as present in the KATRIN setup, one groups the coils of equal orientation and performs the appropriated coordinate transformation of the field.

Fully three dimensional magnetic field calculation As mentioned, at KATRIN not only axially symmetric magnetic fields are present. For certain background processes (see Chap. 5) even tiny deviations from perfect axial symmetry are of great importance. For this reason, fully three dimensional magnetic field calculation methods are needed.

Integrated Biot Savart A coil of arbitrary shape can be made up of small current segments. The magnetic field $\vec{B}$ of a current segment is given by the integrated Biot Savart law

$$\vec{B}(\vec{r}) = \frac{\mu_0}{4\pi}\vec{L} \times \vec{I}, \tag{4.16}$$

$$= \frac{\mu_0}{4\pi}\left(\frac{\hat{r}_1 + \hat{r}_2}{R+l} + \frac{\hat{r}_1 + \hat{r}_2}{R-l}\right) \times \vec{I}, \tag{4.17}$$

where $\hat{r_1}$ and $\hat{r_2}$ are unit vectors pointing from the field point to the begin and end position of the current segment, respectively. The parameter R is given by $R = |\vec{r_1}| + |\vec{r_2}|$ and the length of the segment is $l = |\vec{r}_2 - \vec{r}_1|$. The parameter μ_0 denotes the permeability. By adding up the field of the single current segments the field of arbitrarily shaped coils, for instance the non-perfectly circular LFCS and the EMCS elements, can be computed.

Magnetic dipoles Another source of non-axially symmetric field components are magnetized steel bars in the reinforced concrete building parts. These can be approximated as magnetic dipoles. To compute the field of a magnetic dipole two magnetic charges are assumed to sit at the ends of the dipole. The magnetic charge Q is given by the magnetization $\vec{m}$ of the bar and its radius r_{bar}

$$Q = (\vec{m} \cdot \vec{n})\,\pi r_{\text{bar}}^2, \tag{4.18}$$

where $\vec{n}$ is the normal vector to the cross section of the bar, and, correspondingly, $\vec{m} \cdot \vec{n}$ is the magnetic charge density. The magnetic field H is then given by

$$\vec{H}(\vec{r}) = Q\frac{\mu_0}{4\pi}\left(\frac{\vec{r}-\vec{r}_1}{(\vec{r}-\vec{r}_1)^3} - \frac{\vec{r}-\vec{r}_2}{(\vec{r}-\vec{r}_2)^3}\right), \quad (4.19)$$

where $\vec{r_1}$ ($\vec{r_2}$) denote the start (end) position of the dipole, while μ_0 denotes the permeability.

Electric field calculation

Unlike the case of the magnetic field calculation, the sources of the electric field, i.e. the electric charge densities are not known at the beginning of the simulation. The only known quantity is the electric potential on the electrodes. Consequently, as a first step the charge densities need to be computed. At KATRIN very large dimensions of the order of tens of meters (main spectrometer) concur with very small dimensions of the order of μm (e.g. wire electrode). The most suitable method for coping with large dimensional differences (at KATRIN: factor of 10^5) is the Boundary Element Method (BEM).

Boundary element method At first the electrode geometry is sub-divided into sub-elements. In case of axially-symmetric situations these sub-elements can be ring like structures, whereas for fully arbitrary shapes a discretization into triangles and rectangles is used.

The potential is a linear function of the charge densities of all sub-elements

$$U_i(\vec{r}) = \sum_{j=1}^{N} C_{ij}(\vec{r})\sigma_j, \quad (4.20)$$

with $C_{ij} = C_j(\vec{r}_i)$ being the so-called Coulomb-matrix-elements which are defined as

$$C_j(\vec{r}_i) = \frac{1}{4\pi\varepsilon_0}\int_{S_j}\frac{1}{|\vec{r}_i - \vec{r}_S|}\mathrm{d}^2\vec{r}_S. \quad (4.21)$$

First, the purely geometrical Coulomb matrix elements are calculated. Then the system of linear equations is solved by either the Gauss-Jordan-algorithm, or the Lower Upper (LU) method.

If the system has too many elements (more than 10,000) these methods are no longer applicable due to insufficient working memory of the computer. In these cases iterative solving methods are used. KASSIOPEIA provides the Gauss-Seidel method and the newly developed Robin Hood method [19], which was very successfully tested for the detector region of KATRIN [20].

Axially symmetric electric field calculation In a manner very similar to the magnetic field calculation methods, elliptic integrals and Legendre polynomial expansion are used for electric field and potential calculation [1]. The electric field and potential of an infinitesimally thin cone (ring) can be expressed in terms of elliptic integrals. To go from a ring to a cone of realistic dimensions, one has to integrate along the extension of the cone. As in the case of the magnetic field calculation, the computation time can be drastically reduced by using Legendre polynomial expansion, in a very similar fashion, as explained above.

Fully three dimensional electric field calculation The entire inner surface of the main spectrometer is covered by a wire electrode system (see Fig. 3.8). In a region close to the wires the fields can no longer be assumed to be axially symmetric. In this case a complete simulation of individual wires and their holding structure has to be performed. Accordingly, wire electrodes are approximated by line segments of constant charge. Full electrodes are discretized in triangles and rectangles and the potential can be computed by using Coulomb integration.

These methods require extreme computing power, therefore a huge effort is currently put into further developments allowing for faster three dimensional electric field calculations. These attempts are focused on parallel computation algorithms and usage of GPUs. And on the other hand, also new methods based on multi-pole expansion are pursued [20].

However, in most simulations the particles are propagating far enough from the electrodes so that axial symmetry is a sufficiently good approximation.

4.3.4 Particle Detection (KESS)

KATRIN Electron Scattering in Silicon (KESS) is an event-by-event simulation of electrons in silicon [21]. It has been especially designed for low-energy electrons with kinetic energies below 50 keV, which is the energy range of interest for KATRIN and is not applicable by GEANT4, even in its low energy extension.

The processes included in KESS are thus: elastic scattering [22], inelastic scattering [23–25], ionization [26], atomic relaxation [27] and the vacuum-silicon transition [28] probabilities.

4.3.5 Source Spectrum Calculation

The Source Spectrum Calculator (SSC) provides a realistic source model for the event generators of KPAGE [29]. The module implements the differential β-spectrum of tritium according to Fermi's theory, where the following corrections are taken into account:

- Radiative corrections [30]
- Tritium purity, measured by laser Raman spectroscopy
- Theoretical final state distributions [31, 32]
- Doppler broadening.

In addition to being a tool for the event generators, SSC can be used for direct systematic analysis. In this case the code creates integrated spectra, while approximatively including source properties like the velocity, density and temperature profile of molecules [33, 34], as well as energy loss due to scattering and synchrotron radiation, and finally the spectrometer resolution and detector effects [35].

4.4 User Interface

The KASSIOPEIA configuration follows an intuitive approach, which can easily be handled by a user intending to run a simulation. Its structure complies to following basic questions in order:

- How many Events are to be simulated? (Run Configuration)
- Which kind of Events are to be simulated? (Event Configuration)
- Where does your simulation take place? (Track Configuration)
- What type of simulation algorithm should be applied to the Tracks? (Step Configuration)
- Which electro-magnetic fields are present? (Field Configuration).

To define the actual configuration the user first has to equip so-called Toolboxes with the necessary tools for the simulation, e.g. Generators, Step Strategy tools, etc. Following that, the instructions of how to use the tools throughout the simulation are given in the so-called KASSIOPEIA configuration file.

4.4.1 Toolbox Configuration

KASSIOPEIA comprises 5 main Toolboxes, corresponding to the different physical Modules of the software:

- Generator Toolbox
- Step strategy Toolbox
- Field Toolbox
- Geometry Toolbox
- Source Spectrum Toolbox.

In the Generator Toolbox all generators, which might possibly be needed are configured. The Step Strategy Toolbox comprises the configuration of KTRACK and KESS. Here all required Processes, Step Sizes and Exit Conditions are configured. In the

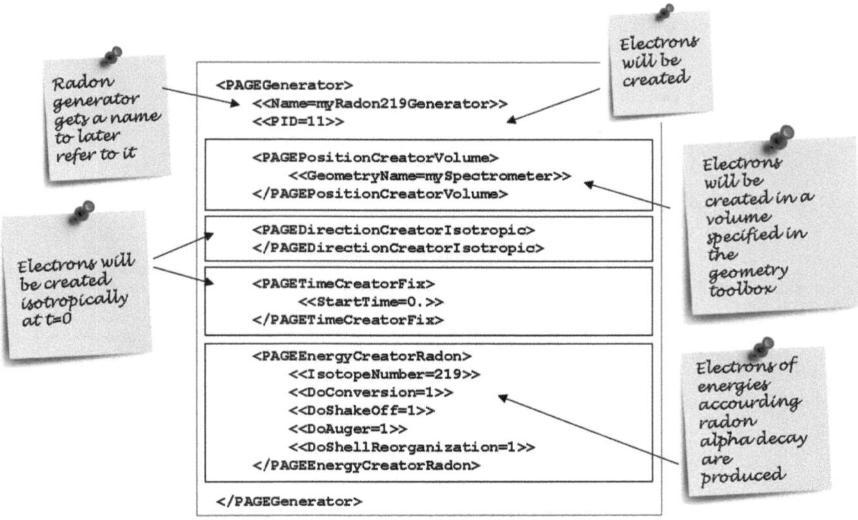

Fig. 4.6 Part of generator toolbox configuration file

Field Toolbox, the fields which are present in the setup are configured. The Geometry Toolbox contains geometrical shapes. These shapes are referred to for particle creation in a volume or on a surface, for navigation, and exit conditions like "geometry hit". In the Source Spectrum Toolbox, the tritium spectrum and corrections to it can be configured.

When this configuration file is read, the Initializer/Builder system of KASSIOPEIA actually instantiates and configures the Modules described, and then registers them inside the Toolbox that the system is set to work with. Other pieces of KASSIOPEIA can then query the Toolbox to retrieve these prepared Modules.

An example for the configuration of a ^{219}Rn Event Generator in the Generator Toolbox is shown in Fig. 4.6.

4.4.2 Instruction Configuration

As mentioned above, KASSIOPEIA has the ability to change the configuration depending on the position of the particle. Correspondingly, a particle can change its behavior once it enters a specific Region. For instance, the behavior of the particle will completely change when it enters the silicon detector. Another example would be that the user might reduce the step size when the particle enters an interesting area. Regions are defined by referring to objects in the Geometry Toolbox. The definition of a certain behavior in a Region occurs in the instruction (KASSIOPEIA) file. An examples of how this file is edited is shown in Fig. 4.7.

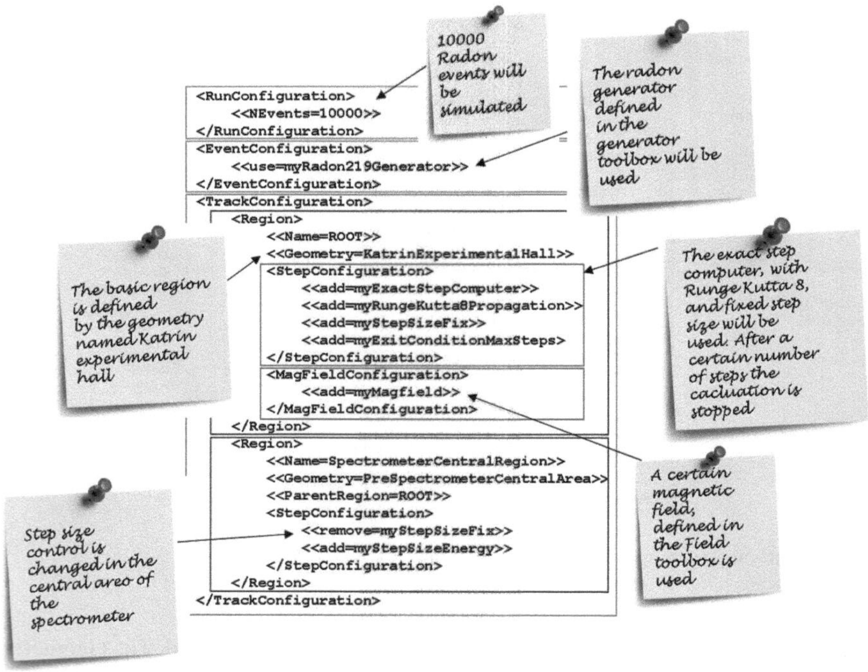

Fig. 4.7 Kassiopeia configuration file

4.5 Output

The standard format for all output data is the ROOT TTree. The output TTrees are stored in a TFile. They consist of three trees, one for event, one for track and one for step data. A standard tool for analyzing these data is available. Furthermore, KASSIOPEIA provides tools for graphical display of the data.

The program package KASSIOPEIA thus represents an extremely versatile and powerful tool to investigate specific systematic effects, to validate models of background processes and to optimize the electromagnetic layout of KATRIN. The following discussions of cosmic muon induced background (Chap. 5), Penning traps (Chap. 6), background from nuclear decays (Chap. 7) and background reduction by the ECR technique (Chap. 8) would not have been possible without this unique code package.

References

1. F. Glück, Axisymmetric electric field calculation with zonal harmonic expansion. Prog. Electromagnet. Res. B **32**, 319–350 (2011)
2. F. Glück, Axisymmetric magnetic field calculation with zonal harmonic expansion. Prog. Electromagnet. Res. B **32**, 351–388 (2011)

3. W. Käfer, Investigation of the KATRIN sensitivity. PhD thesis, KIT, 2012
4. E. Browne, Nuclear data sheets for A = 215, 219, 223, 227, 231. Nucl. Data Sheets **93**(4), 763–1061 (2001)
5. S.C. Wu, Nuclear data sheets for A = 216. Nucl. Data Sheets **108**(5), 1057–1092 (2007)
6. M.S. Rapaport, F. Asaro, I. Perlman, K-shell electron shake-off accompanying alpha decay. Phys. Rev. C **11**, 1740–1745 (1975)
7. M.S. Rapaport, F. Asaro, I. Perlman, M- and L-shell electron shake-off accompanying alpha decay. Phys. Rev. C **11**, 1746–1754 (1975)
8. M.S. Freedman, in *Ionization by Nuclear Transitions*, Conference: Summer course in atomic physics, Carry-le-Rouet, France, 31 Aug 1975; Other Information: Orig. Receipt Date: 30-JUN-76, p. 18, Jan 1975
9. J. Baró, J. Sempau, J. Fernández-Varea, F. Salvat, PENELOPE: an algorithm for Monte Carlo simulation of the penetration and energy loss of electrons and positrons in matter. Nucl. Instrum. Methods Phys. Res., Sect. B **100**(1), 31–46 (1995)
10. PENELOPE homepage, http://www.oecd-nea.org/tools/abstract/detail/nea-1525. Last update: 23 May 2011
11. S. Szucs, J.M. Delfosse, Charge Spectrum of Recoiling ^{216} Po in the α-Decay of ^{220}Rn. Phys. Rev. Lett. **15**, 163–165 (1965)
12. J.S. Hansen, Internal ionization during alpha decay: a new theoretical approach. Phys. Rev. A **9**, 40–43 (1974)
13. F. Glück, Runge-Kutta method for numerical solution of differential equation system. http://fuzzy.fzk.de/bscw/bscw.cgi/d479152/rungekutta.pdf
14. P.W. Sharp, Numerical comparisons of some explicit Runge-Kutta pairs of orders 4 through 8. ACM Trans. Math. Softw. **17**, 387–409 (1991)
15. J. Barrett, Embedded Runge-Kutta Steppers for KMath. http://fuzzy.fzk.de/bscw/bscw.cgi/15316
16. S. Filippi, J. Gräf, New Runge Kutta Nystroem formula-pairs of order 8(7), 9(8), 10(9) and 11(10) for differential equations of the form $y'' = f(x, y)$. J. Comput. Appl. Math. **14**(3), 361–370 (1986)
17. J.C. Butcher, *Numerical Methods for Ordinary Differential Equations* (Wiley, Ltd., 2005)
18. W.H. Press, S.A. Teukolsky, W.T. Vetterling, B.P. Flannery, *Numerical Recipes 3rd Edition: The Art of Scientific Computing*, 3rd edn. (Cambridge University Press, New York, 2007)
19. J. Formaggio, *RobinHood - A new Electrostatics Solver*. KATRIN Collaboration meeting, Oct 2010
20. J. Formaggio, *RobinHood, FFTMs, GPUs.. oh MY*. KATRIN Collaboration meeting, Mar 2011
21. P. Renschler, KESS - A new Monte Carlo simulation code for low-energy electron interactions in silicon detectors. PhD thesis, KIT, 2011
22. F. Salvat, A. Jablonski, C.J. Powell, elsepa-Dirac partial-wave calculation of elastic scattering of electrons and positrons by atoms, positive ions and molecules. Comput. Phys. Commun. **165**(2), 157–190 (2005)
23. D.R. Penn, Electron mean-free-path calculations using a model dielectric function. Phys. Rev. B **35**, 482–486 (1987)
24. H. Bichsel, Straggling in thin silicon detectors. Rev. Mod. Phys. **60**, 663–699 (1988)
25. R.C. Alig, S. Bloom, C.W. Struck, Scattering by ionization and phonon emission in semiconductors. Phys. Rev. B **22**, 5565–5582 (1980)
26. D.T. Cromer, D. Liberman, LASL report, vol. 4403, 1970
27. G. Fraser, A. Abbey, A. Holland, K. McCarthy, A. Owens, A. Wells, The X-ray energy response of silicon Part A. Theory. Nucl. Instrum. Methods Phys. Res., Sect. A **350**(1–2), 368–378 (1994)
28. J.M. Fernández-Varea, X. Llovet, F. Salvat, Cross sections for electron interactions in condensed matter. Surf. Interface Anal. **37**, 824–832 (2005)
29. M. Hötzel, *Berechnung von KATRIN Messspektren unter Einbeziehung der fensterlosen gasförmigen Tritiumquelle*. Master's thesis, KIT, 2009
30. W. Repko, C. Wu, Radiative corrections to the end point of the tritium β-decay spectrum. Phys. Rev. **C28**(1983), 2433 (1983)

31. N. Doss et al., Molecular effects in investigations of tritium molecule β-decay endpoint experiments. Phys. Rev. C **73**(2006), 025502 (2006)
32. N. Doss, J. Tennyson, Excitations to the electronic continuum of $3HeT^+$ in investigations of the T_2 beta-decay experiments. J. Phys. B: At. Mol. Opt. Phys **41**(2008), 125701 (2008)
33. F. Sharipov, *Numerical calculation of tritium flow through the KATRIN beam line*. KATRIN, Report, 2003
34. F. Sharipov, "*Influence of temperature variations and acoustic waves on the column density. Calculations of the velocity distribution function*". KATRIN Report, 2009
35. K. Eitel et al., *Estimate of the KATRIN sensitivity on the neutrino mass*. KATRIN, Report, 2003

Chapter 5
Muon Induced Background

Cosmic ray-induced background processes are of major concern for all low count rate experiments. While $0\nu\beta\beta$ and dark matter searches are performed in underground laboratories, the KATRIN spectrometer is exposed to the full flux of cosmic muons. Every second, about 190 cosmic muons with energies of $E_{\mu} > 1$ GeV patter down onto each square meter of the KATRIN main spectrometer. The muons easily penetrate the 3 cm thick stainless steel vessel and produce electrons on the inner surface. In an electromagnetic cascade, electrons are created over a large energy range. While the energy spectrum peaks at low energies, its high-energy tail extends to about 1 MeV. One expects approximately 10^5 electrons/s being produced at the surface.

In this chapter, the planned countermeasures against cosmic muon-induced background are presented. In particular, the impact of non-axially symmetric magnetic fields on the background production mechanism will be discussed.

5.1 Electric and Magnetic Shielding

Two safeguards preventing μ-induced secondary electrons from reaching the detector are available at KATRIN: magnetic and electric shielding.

In the main spectrometer two superconducting solenoids at the entrance and exit and an air coil system produce a magnetic field of about $B^{e} = 4.5$ T at the entrance/exit and $B^{c} = 3 \cdot 10^{-4}$ T in the center. The magnetic field lines run approximately parallel to the main spectrometer wall. For this reference configuration the field lines ending at the outer radius of the detector, have a distance of about 30 cm to the wire electrode (see Fig. 5.1).

Electrons from the wall and wire electrode are typically very low in energy (with negligible magnetic rigidity) and are thus reflected back to the surface by the Lorentz force constraining them on a cyclotron motion around the magnetic field lines (see Fig. 5.1). It is only in areas where the field lines are not parallel to the tank that

S. Mertens, *Background Processes in the Electrostatic Spectrometers of the KATRIN Experiment*, Springer Theses, DOI: 10.1007/978-3-319-01177-6_5,

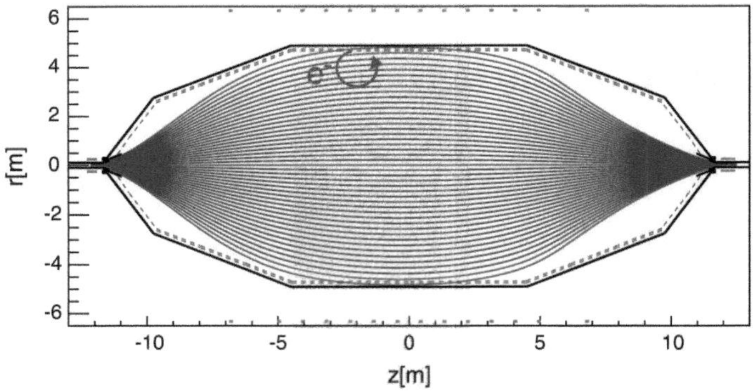

Fig. 5.1 Working principle of the magnetic shielding. The plot shows a cross section of the main spectrometer. In *green* the superconducting magnets and the air coils system is displayed. The *dashed purple line* represents the wire electrode. (The wires run parallel to the beam axis). The *red arrow* illustrates how electrons created by cosmic muons are reflected back to the vessel hull by the magnetic field. The radius of cyclotron motion is artificially increased to make the motion more visible. The *red field lines* are not connected to the detector. The minimal distance of the *blue field lines* to the wire electrode is about 30 cm. The detector is therefore further shielded from electrons from the wall and wires

electrons starting under a small angle to the surface, can circumvent the magnetic reflection. Depending on their starting energy and angle three different behaviors can be distinguished:

- The electrons leave the spectrometer, however, they are guided along field lines that do not end at the detector, but pass by it.
- Observations in measurements at Mainz and the pre-spectrometer have revealed that some electrons can, due to a non-adiabatic motion, directly reach the detector. This phenomenon, however, will not be discussed in this work.
- An electron moving from the central region of the spectrometer (at low magnetic field) towards the entrance/exit (at high magnetic field) can be reflected by the magnetic mirror effect and be stored. As will be explained in the following, these electrons are the prime candidates to produce background.

Based on previous measurements at the Mainz spectrometer, the expected magnetic shielding factor at the main spectrometer is of the order of $r^{\mathrm{mag}} = 10^5$.

To further shield the spectrometer from low energy secondary electrons, a two-layer wire electrode is installed at a distance of 15 cm (inner layer) and 22 cm (outer layer) to the surface of the main spectrometer vessel (see Fig. 3.8). By setting the wires to a more negative potential $\Delta \mathrm{V} = -100$ V(-200 V) for outer (inner) layer than the tank, low-energy electrons are repelled back electro-statically to the vessel wall as visualized in Fig. 5.2. The additional reduction factor is expected to be of the order of $r^{\mathrm{el}} = 10^2$, again based on measurements at the Mainz spectrometer.

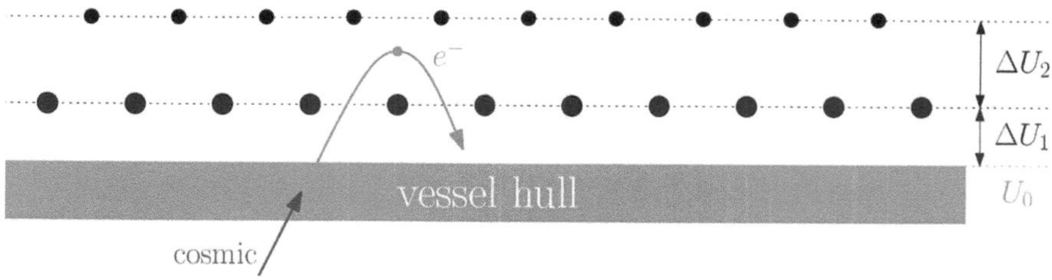

Fig. 5.2 Working principle of wire electrode. Here a cross section of a small part of the main spectrometer vessel perpendicular to the wires is shown. The plot illustrates how electrons created by cosmic muons are reflected back to the vessel hull by the more negative potential of the wire electrode

With $r^{\mathrm{mag}} = 10^5$ and $r^{\mathrm{el}} = 10^2$ as well as the expected μ-induced secondary electron rate of $10^5\ \mathrm{s}^{-1}$, a first straightforward estimation of the background rate of $\mathrm{b} = 10^{-2}\ \mathrm{s}^{-1}$ can be made.

5.2 Circumvention of Shielding and Background Production Mechanism

Simulations with GEANT4 [1, 2] and Penelope [3] revealed that about 2,000 electrons per second are produced in an energy range from 1–50 keV over the entire inner surface of the main spectrometer [4]. These high-energy electrons are not repelled by the wire electrode and can, as mentioned, be stored in the volume of the spectrometer or due to their large cyclotron radius traverse the sensitive spectrometer volume. Furthermore, electrons of all energies directly produced on the wire surface can enter the main spectrometer and be stored there. In a perfectly axially-symmetric field, the stored electrons are situated on a surface that lies outside of the sensitive volume, where they cannot produce any background rate. Following two processes, however, lead to background production:

- At first, non-axially symmetric field components lead to a radial drift that allows the stored electrons to enter the sensitive volume of the spectrometer.
- In this sensitive area, the stored electrons can produce secondary (in our case tertiary) electrons via ionization of residual gas molecules. These secondaries are eventually released from the trap, reach the detector and finally produce a background rate.

In Chap. 7 the background production mechanism of stored electrons will be discussed in detail, in this chapter, however, the focus is on the radial drift due to non-axially symmetric field components. A radial drift can be caused by two generic mechanisms: the $\vec{E} \times \vec{B}$ and the $\vec{\nabla}|\vec{B}| \times |\vec{B}|$ drift:

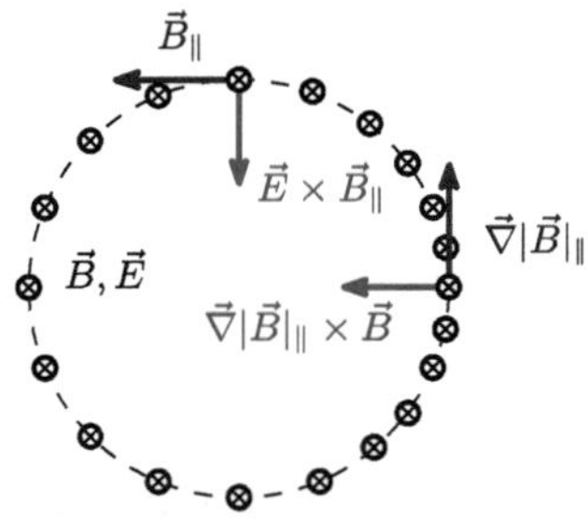

Fig. 5.3 Radial drift motion. This plot visualizes the occurrence of radial drift motion. The figure shows a cross section of a flux tube of the main spectrometer at an arbitrary axial position. The electric and magnetic fields point into the plane. A magnetic field component oriented azimuthally to the flux tube $\vec{B}_{\|}$ leads to a radial $\vec{E} \times \vec{B}$-drift. Accordingly, a gradient of the magnetic field in azimuthal direction $\vec{\nabla}|\vec{B}|_{\|}$ produces a radial $\vec{\nabla}|\vec{B}| \times |\vec{B}|$-drift

$$v_{\vec{\nabla}|\vec{B}|\times\vec{B}} = \frac{\gamma m}{q|B|^3}\frac{v_{\|}^2 v_{\perp}^2}{2}\left(\vec{B} \times \nabla|\vec{B}|\right), \tag{5.1}$$

$$\vec{v}_{\vec{E}\times\vec{B}} = \vec{E} \times \vec{B}\left(\frac{1}{\vec{B}^2}\right), \tag{5.2}$$

$$\vec{v}_{\text{drift}} = \vec{v}_{\vec{\nabla}|\vec{B}|\times\vec{B}} + \vec{v}_{\vec{E}\times\vec{B}}. \tag{5.3}$$

The presence of an azimuthal magnetic field component $\vec{B}_{\|}$ or a magnetic field gradient in azimuthal direction $\vec{\nabla}|\vec{B}|_{\|}$ leads to radial drifts, as visualized in Fig. 5.3. Non-axially symmetric electric fields can also cause radial $\vec{E} \times \vec{B}$ drifts, however, they are not discussed in this chapter.

Radial electric field components $\vec{E}_{\perp}$, arising for instance from a non-constant electric potential in the analyzing plane, as well as radial magnetic field gradients $\vec{\nabla}|\vec{B}|_{\perp}$ result in an azimuthal drift of the electron. This azimuthal drift is called magnetron drift motion and is visualized in Fig. 5.4. Since radial fields typically are much larger than azimuthal field components in the KATRIN setup, the magnetron drift is much faster that the radial drift motion.

The radial drift leads to a deformation of the electron path. However, the electron describes a closed path after a full magnetron circulation. Accordingly, if the non-axially symmetric fields did not guide the electron into the sensitive volume within one magnetron turn, the trajectory will always stay outside of the flux tube.

5.3 Sources of Non-axially Symmetric Magnetic Fields

The dominant contribution to the magnetic field in the main spectrometer is generated by the superconducting solenoids (pre-spectrometer magnets: $B_{\text{PS}} = 4.5$ T, pinch magnet: $B_{\text{p}} = 6$ T, detector magnet $B_{\text{d}} = 3 - 6$ T) and the Low Field Coil System

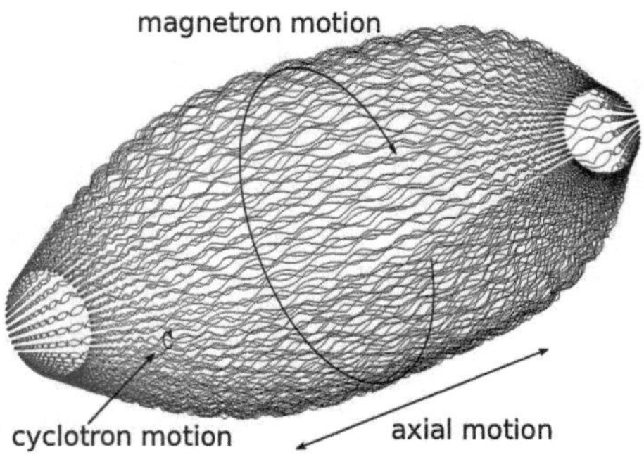

Fig. 5.4 Different components of a stored electron motion in the electro-magnetic field of the main spectrometer. The motion is composed of a cyclotron motion around the magnetic field lines (fastest: $t_{\text{cycl}} \sim 100$ ns) an axial motion along the beam axis (second fastest: $t_{\text{axial}} \sim 10\ \mu$s), and a magnetron drift motion (slowest: $t_{\text{mag}} \sim 100\ \mu$s) caused by $\vec{E} \times \vec{B}$ and $\vec{\nabla}|\vec{B}| \times |\vec{B}|$-drifts, where $\vec{E}$ and $\vec{\nabla}|\vec{B}|$ are oriented in radial direction

(LFCS) of the air coil system. All these magnets are aligned along the beam axis and hence produce an axially symmetric field. Potential sources of non-axial magnetic fields result from

- a tilt of the coils, e.g. magnets of the transport section and non perfectly aligned main spectrometer magnets,
- external magnetic fields, for instance the stray magnetic field of the monitor spectrometer, or magnetic materials from the building structure,
- non-perfectly compensated earth magnetic field,
- deformation of coils, in particular the LFCS.

Table 5.1 shows the quantitative contribution to $\vec{B}_{\|}$ and $\vec{\nabla}|\vec{B}|_{\|}$ of the different sources. In the following the two dominant sources, a potential deformation of the air coil system and magnetic materials are discussed more thoroughly.

5.3.1 Magnetic Materials

In the KATRIN spectrometer hall at the KIT Campus North site, steel bars are embedded in the concrete walls for reinforcement. Close to the main spectrometer

Table 5.1 Main sources of non-axially symmetric field components. The table shows the maximal values of $\vec{B}_{\|}$ and $\vec{\nabla}|\vec{B}|_{\|}$ in the analyzing plane

Source	$\vec{B}_{\|}$[T]	$\vec{\nabla}|\vec{B}|_{\|}$[T/m]
Transport section	$6 \cdot 10^{-9}$	$1 \cdot 10^{-9}$
Tilt of superconducting coils	$2 \cdot 10^{-7}$	$3 \cdot 10^{-9}$
EMCS + earth magnetic field	$6 \cdot 10^{-5}$	$6 \cdot 10^{-9}$
Monitor spectrometer	$1.5 \cdot 10^{-7}$	$1 \cdot 10^{-7}$
Deformation of air coils	$2 \cdot 10^{-7}$	$8 \cdot 10^{-8}$
Magnetic materials	$1 \cdot 10^{-6}$	$6 \cdot 10^{-7}$

Fig. 5.5 Magnetic materials in the KATRIN hall. This *photograph* shows the KATRIN main spectrometer hall when it was built. It is visible that in the central region of the hall stainless steel was used, whereas in the walls and in the floor close to the walls normal steel is embedded into the concrete

stainless steel bars were used, owing to its austenitic nature, whereas further away from the vessel the steel bars are made of normal steel, as visualized in Fig. 5.5. The normal steel bars can create a small non-axially symmetric magnetic field at the position of the main spectrometer. This magnetic field can arise from two cases: firstly, from a remanent magnetization created during construction of the hall, and secondly, the presence of an external magnetic field, such as from the solenoids of KATRIN, causes the steel bars to change their magnetization. Therefore, a permanent magnetic field monitoring system, will be installed in the close vicinity of the air coil system: Each coil will be be equipped with an ultra precise mobile sensor unit that runs along the holding structure [5].

To investigate the influence of the magnetic materials a model was developed, in which the steel bars are approximated as a magnetic dipole array, consisting of about 2,000 dipoles distributed in the area of the wall and floor. To find the magnetization of the magnetic dipoles, more than 1,000 magnetic field measurements at various distances to the wall were performed with a three axis fluxgate field sensor of a precision of $5 \cdot 10^{-8}$ T [6]. The magnetizations were computed by fitting them to the field measurement results [7]. In the future, the dipole model will be permanently adjusted to the measurement results of the magnetic monitoring system. To test the model, the simulated magnetic field was compared to further magnetic field measurements in the volume of the main spectrometer (see Fig. 5.6).

Using this model, the magnetic field causing a radial drift can be computed [8]. To speed up the calculation an interpolation grid was used. Figure 5.7 shows the resulting magnetic field components $\vec{B}_{\parallel}$ and $\vec{\nabla}|\vec{B}|_{\parallel}$ in the analyzing plane.

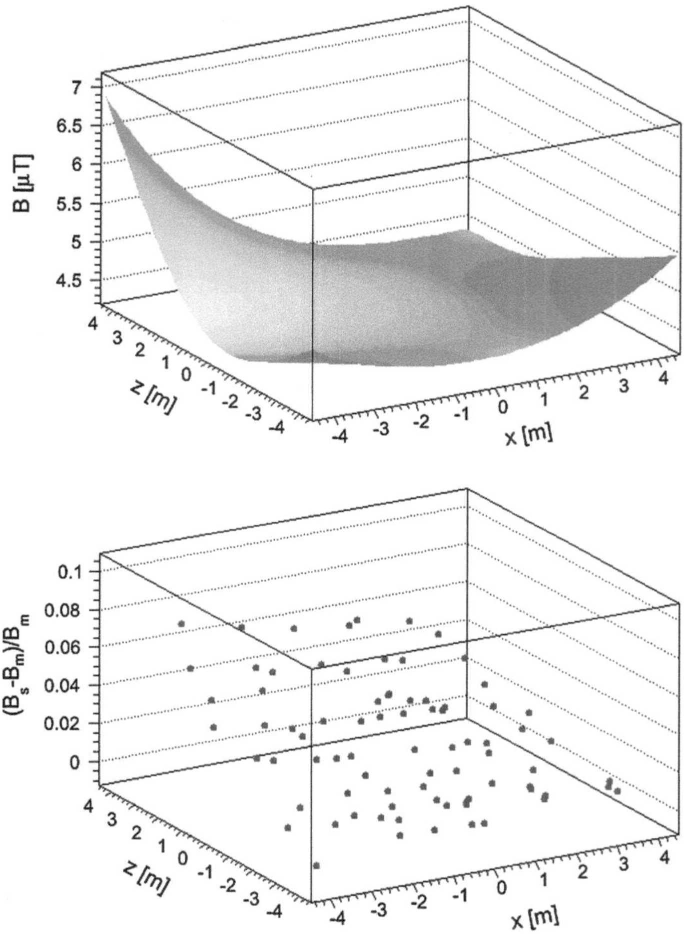

Fig. 5.6 Model of magnetic materials. *Top*: simulated field of magnetic materials in a horizontal plane in the main spectrometer. *Bottom*: comparison of simulated and measured magnetic materials. To compare the model with data, more than 100 measurements of the magnetic field inside the spectrometer were taken with the flux-gate sensor mentioned in the text [6, 7]. The deviation of the simulated B_s and measuremed B_s magnetic field is less than 10% for all measurement points, which is of sufficient accuracy for the purpose of this investigation

5.3.2 Deformation of Air Coil System

The LFCS consists of 16 circular coils of a radius of $r = 6.3$ m surrounding the main spectrometer. The purpose of the LFCS is the fine adjustment of the magnetic field in the center of the spectrometer. Laser tracker based measurements of the holding structure showed that there is a deviation of circularity of the support rings by up to 40 mm as shown in Fig. 5.8. A deviation of this size will lead to non-axially symmetric magnetic field components in the spectrometer.

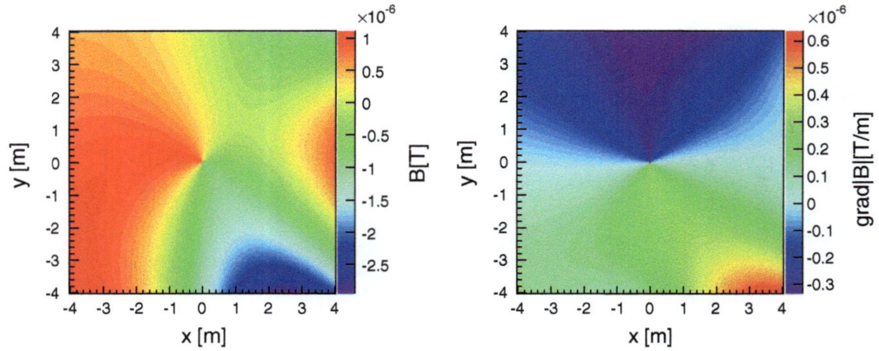

Fig. 5.7 Magnetic field and field gradients due to magnetic materials. The plots show $\vec{B}_{\parallel}$ (*left*) and $\vec{\nabla}|\vec{B}|_{\parallel}$ (*right*) in the analyzing plane

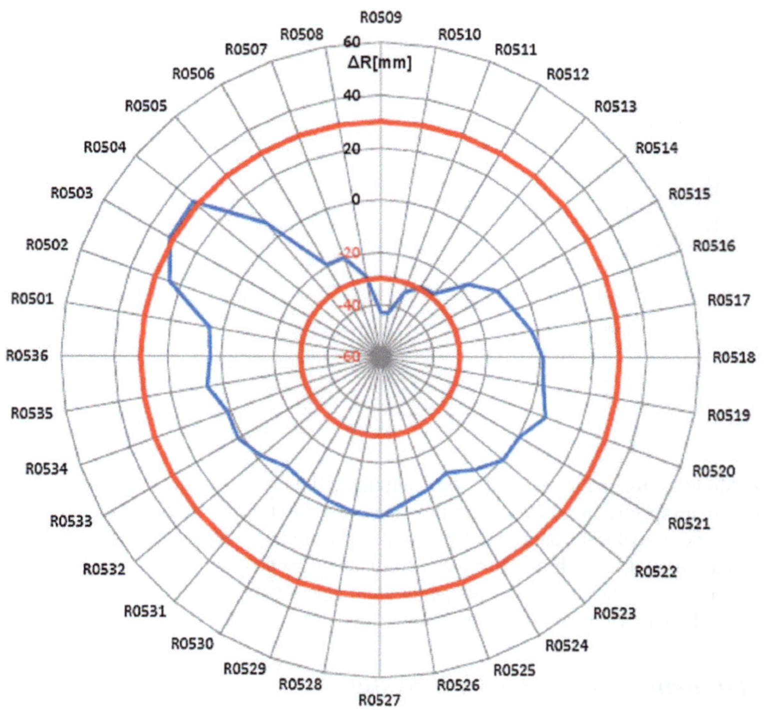

Fig. 5.8 Deformation of low field coil system (LFCS). In this figure the measured deviation of the LFCS coils (*blue line*) from circularity is shown [9]. The *red lines* denote the ±30 mm deviation

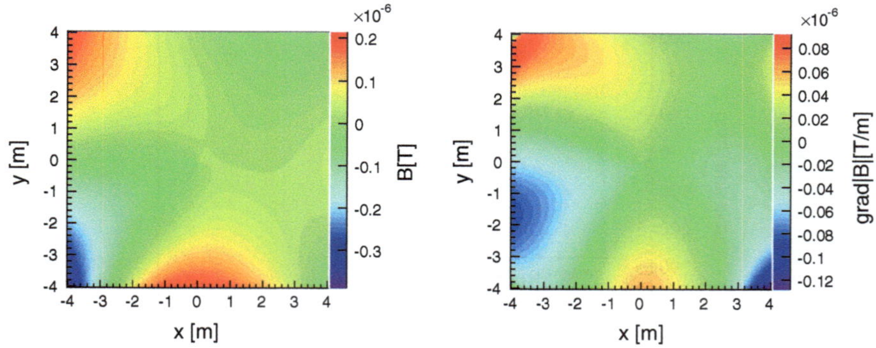

Fig. 5.9 Magnetic field and field gradients due to deformation of LFCS. The plots show $\vec{B}_{\parallel}$ (*left*) and $\vec{\nabla}|\vec{B}|_{\parallel}$ (*right*) in the analyzing plane

To calculate these components a model of the deformed LFCS was developed [8] by dividing the coils up in more than 50,000 separate current segments. The magnetic field of single current segments is given in Eq. (4.16). Non-axially symmetric field components $\vec{B}_{\parallel}$ and $\vec{\nabla}|\vec{B}|_{\parallel}$ arising from the deformation of the LFCS are shown in Fig. 5.9.

5.3.3 *Resulting Radial Drift Velocities*

As evident from Eq. (5.1), $\vec{B}_{\parallel}$ and $\vec{\nabla}|\vec{B}|_{\parallel}$ will lead to radial velocities permitting electrons access to the inner parts of the flux tube. In Fig. 5.10 the radial drift velocities caused by the magnetic material and the deformed LFCS are compared. A radial drift velocity of up to $v_{\text{drift}}^{\text{mm}} = 300$ m/s is caused by the magnetic materials, whereas the deformation of the LFCS only leads to radial drift velocities of $v_{\text{drift}}^{\text{LFCS}} = 100$ m/s.

In the analyzing plane, the electric potential is very homogeneous and hence the electric field is almost zero, therefore, the $\vec{E} \times \vec{B}$-drift vanishes. Only further away from the analyzing plane the electric field is non zero, however, there the magnetic field is larger, which suppresses the radial drift velocity. As shown in Fig. 5.11, the $\vec{E} \times \vec{B}$-drift is already smaller by a factor of about 100 in a region 5 m away from the analyzing plane.

Monte Carlo Simulation of Radial Drift Motion

To investigate the characteristic features of the radial drift motion a Monte Carlo simulation of 83,000 electrons in the main spectrometer in the presence of non-axially symmetric magnetic fields was performed with KASSIOPEIA. The electrons were created on a cylindrical surface close to the wire electrode (length of $l = 9$ m,

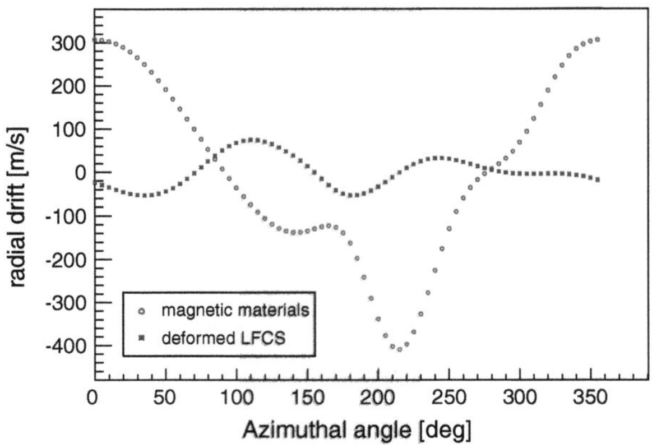

Fig. 5.10 Comparison of drift velocities caused by magnetic materials and the deformed LFCS. This *plot* shows the radial drift velocities on a circle with the radius $r = 3.9$ m in the analyzing plane. Evidently, the LFCS deformation has a much smaller effect than the magnetic materials. Nevertheless, the LFCS coils were corrected by installing wafers at the positions of radii smaller than the nominal radius

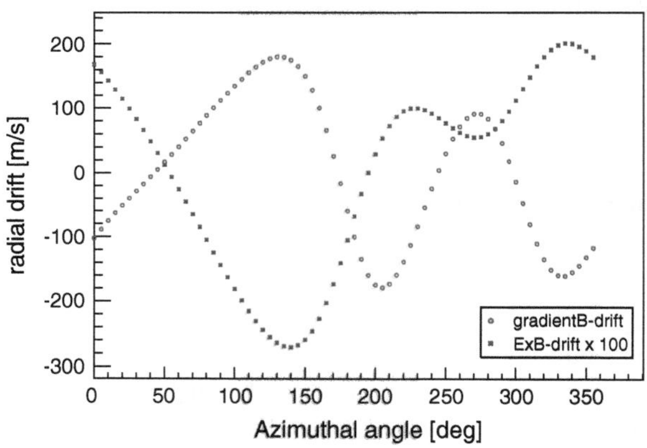

Fig. 5.11 Comparison of $\vec{E} \times \vec{B}$-drift $\vec{\nabla} \times |\vec{B}|$-drift caused by magnetic materials. This *plot* shows the radial drift velocities on a circle with the radius $r = 3.0$ m at $z = 5$ m. The $\vec{E} \times \vec{B}$-drift (here: scaled by a factor of 100) is negligible compared to the $\vec{\nabla}|\vec{B}| \times |\vec{B}|$-drift

radius of r = 4 m) inside of the main spectrometer. Owing to the fact that electrons passing the wire electrodes or starting from a wire have a rather low kinetic energy a flat energy distribution of $1 < E < 100$ eV and an isotropic angular distribution was assumed. The magnetic material model, described in Sect. 5.3.1, was used as a source of a non-axially symmetric magnetic field. The calculation of the trajectory of an electron was terminated as soon as the electron,

- has performed a full magnetron drift
- leaves the spectrometer
- reaches the flux tube (cylindrical volume with r = 3 m, l = 4.5 m).

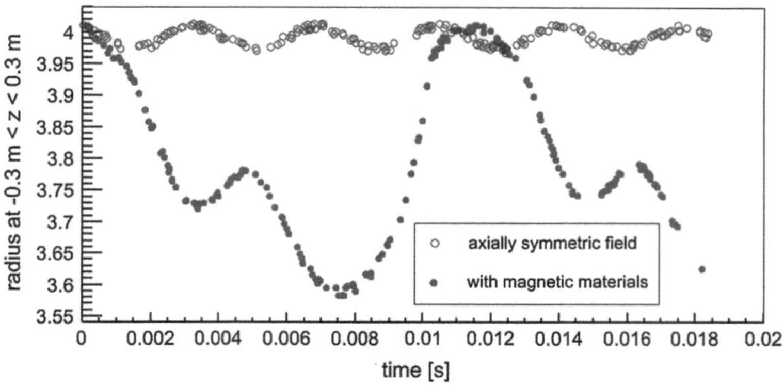

Fig. 5.12 Radial drift motion around the analyzing plane of a stored electron. The figure shows the radial position of a stored electron at $-0.3 < z < 0.3$ m in presence of magnetic materials (*red dots*) and without (*blue dots*). The radial motion of the electron in an axially symmetric field is due to its cyclotron motion. The electron was created at $x = 3.29$ m $y = -2.26$ m and $z = 0.56$ m with an initial kinetic energy of $E = 47.79$ eV.

To speed up the CPU time-consuming simulation only those electrons were simulated, where the azimuthal drift velocity v_{azi} relative to the radial drift velocity v_{r} was low enough to allow for a sizable radial drift within one magnetron turn, i.e. ($\frac{v_{\text{azi}}}{v_{\text{r}}} < 20$).

The simulation reveals that a small fraction of 0.031 % of all electrons can enter the flux tube and create background. Figure 5.12 shows exemplary the corresponding radial positions in the analyzing plane of an electron which reached the flux tube.

5.4 Conclusion

The studies described in this chapter could show that non-axially symmetric magnetic fields will cause radial drift velocities of up to 300 m/s. The strongest source of non-axially symmetric magnetic fields are the magnetic materials used in the floor and walls of the main spectrometer hall. Furthermore, Monte Carlo simulations with KASSIOPEIA revealed that muon-induced stored electrons can reach the sensitive volume of the spectrometer due to these drift processes.

More detailed studies will be carried out in the future which will include the complete calculation of the trajectory of the electrons from the walls through the wire electrode. A full simulation of the expected rate can then be cross-checked with test measurements of the main spectrometer. The interplay of experiment and corresponding Monte Carlo simulation will then allow to optimize the magnetic shielding to minimize cosmic muon induced background.

References

1. S. Agostinelli, J. Allison, Geant4a simulation toolkit. Nucl. Instrum. Methods Phys. Res., Sect. A **506**(3), 250–303 (2003)
2. J. Allison, K. Amako, Geant4 developments and applications. IEEE Trans. Nucl. Sci. **53**, 270–278 (2006)
3. J. Baró, J. Sempau, J. Fernández-Varea, F. Salvat, PENELOPE: An algorithm for Monte Carlo simulation of the penetration and energy loss of electrons and positrons in matter. Nucl. Instrum. Methods Phys. Res., Sect. B **100**(1), 31–46 (1995)
4. B. Jung, Simulation des Myon-induzierten Untergrundes und dessen Speicherwahrscheinlichkeit am KATRIN Hauptspektrometer. Master's thesis, KIT, 2011
5. J. Hergenhahn, Das KATRIN-Luftspulensystem: Messungen des Feldverlaufs und Überwachung mit einem Sensorsystem. Master's thesis, KIT, 2010
6. F. Primdahl, The fluxgate magnetometer. J. Phys. E: Sci. Instrum. **12**(4), 241 (1979)
7. J. Reich, Magnetfeldmessungen und Designarbeiten für das EMCS Luftspulensystem am KATRIN Hauptspektrometer. Master's thesis, KIT, 2009
8. B. Leiber, Non-axially symmetric field and trajectory calculations for the KATRIN-experiment. Master's thesis, KIT, 2009
9. R. Gumbsheimer, *Main Spectrometer - Overview since the last collaboration meeting.* In: Proceedings of KATRIN Collaboration meeting, March, 2010

Chapter 6
Background Due to Penning Traps

Penning traps are a very useful tool in many areas of physics, most notably in atomic physics. They are often used for precision measurements of ion properties, but also for technical applications like vacuum gauges and sputtering technology. However, for KATRIN, Penning traps are an unwelcome guest, most of the time. In the following sections a phenomenological model of the background production mechanism of Penning traps as well as an example of Penning trap induced background observed at the pre-spectrometer will be presented.

6.1 Model of Background Production Mechanism

A Penning trap is based on an interplay of electric and magnetic fields allowing for storage of charged particles. An electric potential well confines the particle in axial direction while a magnetic field binds the particle in radial direction, as visualized in Fig. 6.1. In the case of KATRIN, the electric and magnetic field configurations can lead to a situation where the electric potential along a magnetic field line takes on a minimum, which exactly represents a Penning trap configuration. This most likely occurs close to a complicated geometry, like the inner electrode system.

The stored electrons build up a space charge that counteracts the potential well. Consequently, a Penning trap can be filled only up to a certain equilibrium state. To describe Penning trap induced backgrounds, both filling mechanism and actual background production mechanism need to be understood. Both proceed via messenger particles, which typically are positive ions.

However, within the framework of this thesis this basic understanding was enlarged, as in some cases positive ions can not explain the observed background rate. In those cases photons take on the role of the messenger particles instead. This novel aspect of Penning trap induced background will be illustrated in detail on the basis of dedicated test measurements at the pre-spectrometer.

S. Mertens, *Background Processes in the Electrostatic Spectrometers of the KATRIN Experiment*, Springer Theses, DOI: 10.1007/978-3-319-01177-6_6,

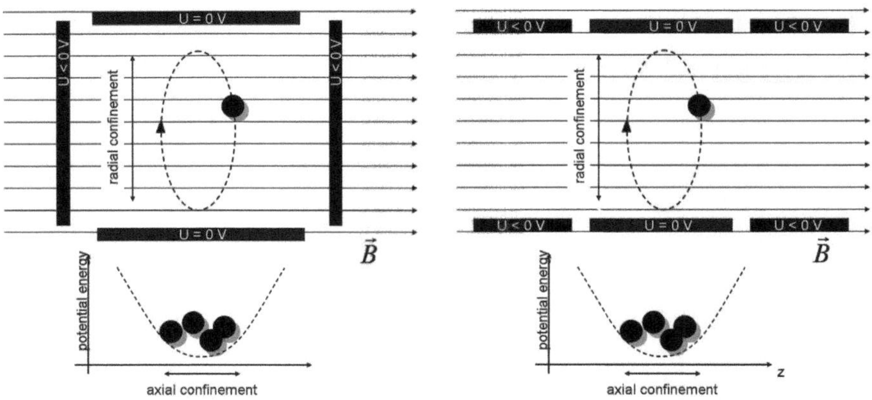

Fig. 6.1 Schematic view of Penning traps. One can distinquish two types of Penning traps: the cathode-cathode Penning trap (*left*), and the vacuum-vacuum Penning trap (*right*). The electron is axially stored due to the electrodes on negative potential, moreover it is radially confined due to the magnetic field oriented in the direction of the axial confinement

6.1.1 Penning Trap Filling Mechanism

To fill a Penning trap one single initial stored electron is needed at least. However, once a small number of electrons are stored in the trap, it fills up exponentially up to a point when space charge effects play a significant role.

Stored electrons produce positive ions via ionization of residual gas molecules [1]. The positive ions are not stored in the electric potential trap, and, furthermore, in a weak magnetic field, their mass can be too large to be confined by the magnetic field lines. Attracted by the negative potential they hit the electrode surface and consequently release electrons that can, in a suitable geometrical situation, fly into the region of the trap. There they have small probability to ionize residual gas. Secondaries produced in the volume of the trap are stored again. This leads to a positive feedback mechanism, i.e. the more electrons are stored the faster the trap is filled.

The filling mechanism can, in principle, also proceed via photons: The kinetic energy of the stored electrons depends on the depth of the trap, which for KATRIN typically is in the keV range. This energy is sufficiently large to excite residual gas molecules and atoms. The subsequent de-excitation produces photons of $E_{\text{photon}} = E_{\text{vac}} - E_{\text{exc}}$, where E_{vac} and E_{exc} denote the energy levels of the vacancy and the level the electron was excited to, as visualized in Fig. 6.2. Furthermore, ionization of residual gas can also lead to photon production. If an electron of an inner shell is removed, photons are created by the subsequent relaxation (see Fig. 6.2). The photon energy is $E_{\text{photon}} = E_{\text{vac}} - E_i$, where E_i denotes the binding energy of the electron filling the vacancy.

The photons are isotropically emitted by the excited or ionized molecules, and, as neutral particles, they are not stored in the trap. When hitting the cathode surface

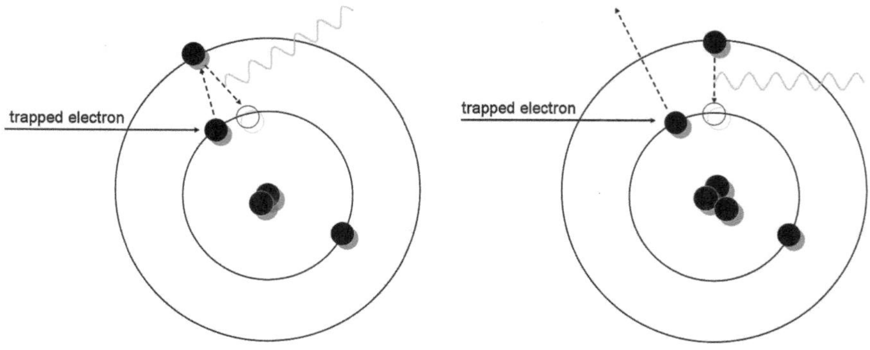

Fig. 6.2 Photon production in Penning trap. The most efficient way to produce photons in Penning traps is via excitation and ionization. The stored electrons are accelerated by the negative potential and gain enough energy to kick out an electron from a deeper shell of a residual gas molecule of atom. Consequently, photons are produced via de-excitation and relaxation

they produce secondary electrons via the photo effect there. These electrons are accelerated back to the trap and can thus create tertiary electrons in the volume of the trap, leading, as in the case of positive ions, to a positive feedback filling mechanism.

6.1.2 Background Production

It is well known that Penning traps cause Penning discharges [2]. Most generally, a Penning discharge can be described as a charge exchange between cathode and anode. For instance, positive ions from the trap can be collected at the cathode registering a small current (weak discharge). On the other hand, it is possible that current from the high voltage electrode leaks through the plasma of stored electrons to the ground electrode (strong discharge). Strong discharges can directly be measured by observing large leakage currents. In the most severe cases, a Penning discharge can lead to an electric breakdown. However, even with very weak discharge ($I_{\text{leak}} < 1$ mA), a Penning trap can still produce large background rates.

As the stored electrons themselves cannot reach the detector, the background production proceeds again via messenger particles, which again, as in the case of the filling mechanism, are photons or positive ions.

Depending on the geometry and the position of the Penning trap, it is the positive ions, produced in the Penning trap, that can reach the sensitive volume of the spectrometer. There, they can create electrons via ionization of residual gas molecules. The cross section for $H_2^+ + H_2 \rightarrow e(\text{total}) \approx 10^{-17}$ cm^2 for a 1 keV ion is rather large and further increasing with energy [3].

Also photons can take on the same role as the positive ions: They can illuminate the sensitive volume and produce secondary electrons via photo-ionization [4]. Photons with an energy of about 14 eV have a cross section for photo-ionization of atomic hydrogen of $\sigma \approx 5 \cdot 10^{-18}\,\text{cm}^2$.

The secondary electrons, when created in the sensitive volume, have a free path to the detector. First produced with low energies, they are accelerated to tank potential on their way to the detector where they are detected in the energy region of interest (i.e. at the so called full U_0 peak).

6.2 Small Penning Trap at the Pre-spectrometer

In this section a series of measurements at the pre-spectrometer, which demonstrate that the existence of small Penning traps can lead to tremendous background rates of kHz, is discussed [5, 6]. In particular, the hypothesis of background production via photons will be confirmed.

6.2.1 Overview of Measurement Series

When the magnetic field at the pre-spectrometer set-up was ramped up to more than 2 T (corresponding to 2 μT in the analyzing plane) a background of up to several hundred Hz was observed. Due to this correlation this background was called the "High-B-Field Background". In the following a series of three measurements, each based on a specific (different) electrode configuration will be discussed. The geometry of the pre-spectrometer, and in particular the labelling of electrodes and magnets is given in Fig. 6.3, while the three different electrode designs are compared in Fig. 6.4.

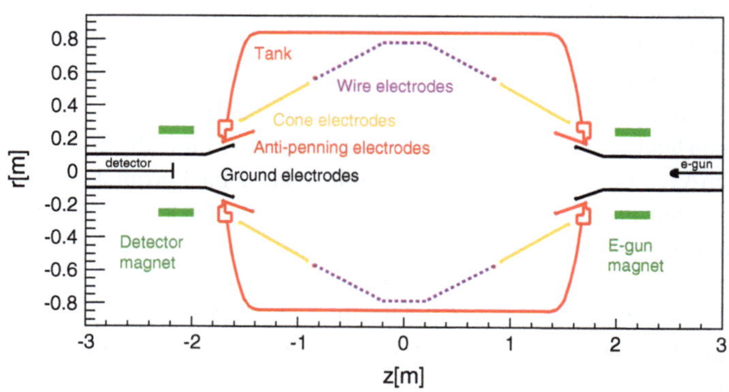

Fig. 6.3 Pre-spectrometer setup. The different colors denote the electrodes that can be set on different voltages. The e-gun is a monochromatic electron source, its tip is made of gold, where electrons are knocked out of its surface by a UV-light source irradiating the tip from the inside. The angle of the e-gun to the beam axis is adjustable. It is set to a variable voltage to produce electrons of different energies. A segmented silicon detector with 8 · 8 pixels (64PD) is mounted on a manipulator, which can be used to displace the detector in x,y and z direction

Fig. 6.4 Comparison of three electrode designs. This photograph shows three different ground electrode designs. The electrode design had a huge impact on the background rate. The electrodes (*from left to right*) correspond to the measurement I-III described in this section. The corresponding simulation of Penning traps for the three geometries can be found in Fig. 6.8

Setup I: Original Electrode Configuration

The "High-B-Field background" was discovered already at the end of 2007. It was observed that when operating at a magnetic field B $>$ 2 T, the rate increased after about 15 min from 100 mHz to 30 Hz on a short time scale of 1 min, as shown in Fig. 6.5.

For a long time a positive ion trap in the center of the spectrometer, that is inherently present in MAC-E-Filters, was under suspicion of causing this background. However, an asymmetric magnetic field measurement revealed that the source of the

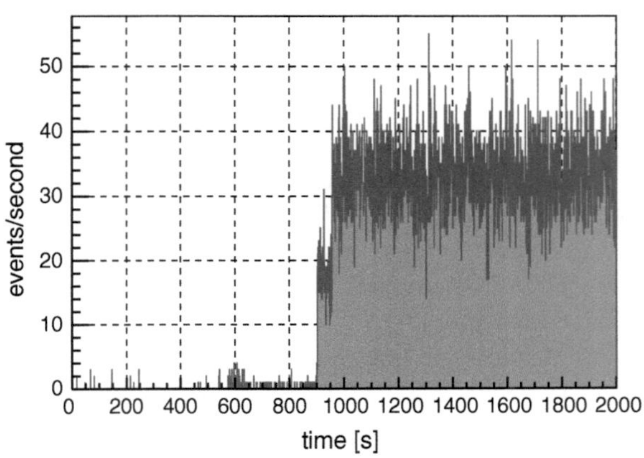

Fig. 6.5 Ingnition of background due to penning trap. Measurement I at pre-spectrometer. Settings: Both magnets are set to B = 3.4 T, all electrodes are on $U = -18$ kV. The background rate increases from about 100 mHz to about 30 Hz in the region of interest (15–21 keV) after 900 s

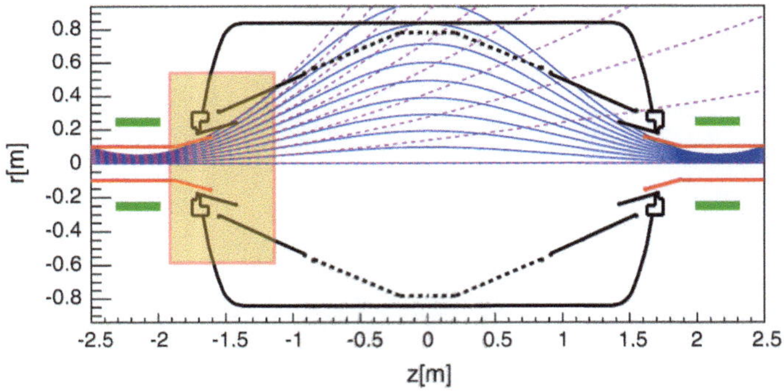

Fig. 6.6 Magnetic field lines in the pre-spectrometer for different magnetic field settings. The *blue lines* represent the magnetic field lines for a symmetric magnetic field setting, whereas the *dashed violet lines* show the field lines for an asymmetric magnetic field configuration ($B_{\text{det}} = 4.5$ T, $B_{\text{e-gun}} = 0$ T). In the asymmetric setting a Penning trap cannot form in the center of the spectrometer. The observation that the background rate did not disappear for an asymmetric magnetic field could prove that the source of background must be located in the entrance/exit region of the pre-spectrometer, where the field lines are identical for both configurations, as indicated by the *yellow area*

background must be located in the entrance/exit region of the spectrometer. In this asymmetric field configuration the detector magnet is set to $B^{\text{det}} = 4.5$ T, whereas the e-gun magnet is not ramped up, i.e. $B^{\text{e-gun}} = 0$ T (see Fig. 6.6). In this magnetic field configuration a trap cannot be present in the center of the spectrometer. Surprisingly, and in contradiction to the ion trap hypothesis, a background rate of about half the initial strength was still measured. Even more striking was the fact that the same characteristic delay of the rate was still observed. Consequently, one could conclude that the source of background is "hiding" at the entrance/exit region.

Setup II: Modified Electrode Configuration

Based on these observations, a new design of the ground electrode was developed and installed [5]. This new design avoided a small electron Penning trap just *below* the end-ring of the ground electrode of $U_{\text{trap}} = 0.6$ kV. Additionally, a wire comb was installed to avoid a small ion trap in the entrance/exit region (see Fig. 6.7).

Surprisingly, the "High-B-Field background" did not decrease, as expected, but did increase to about 6 kHz after this modification.

This observation left only one conclusion: A Penning trap, situated *above* the ground electrode end ring with $U_{\text{trap}} = 0.9$ kV must cause the background rate. This Penning trap was not removed with the redesign of the electrode, but rather increased to a larger value of $U_{\text{trap}} = 2.5$ kV (see Fig. 6.8). The reason for not considering this trap in the first place was that no mechanism was known at that time which could explain the background production of a Penning trap located outside of the electrode system.

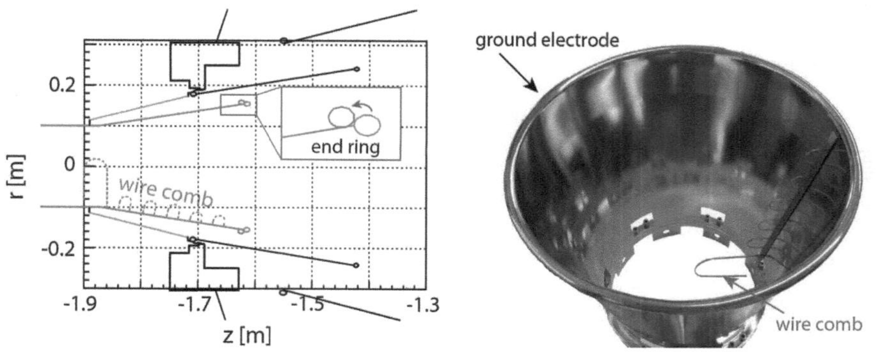

Fig. 6.7 Design features of the 2nd electrode [5]. The end ring of the ground electrode was modified to exhibit a larger radius, to avoid a Penning trap below the end ring. Additionally, a wire comb was installed inside of the ground electrode, to avoid an ion trap in the ground electrode region

Setup III: Final Improved Electrode Configuration

Following these considerations, a final modification to the ground electrode was performed, intended to eliminate all Penning traps (in particular the Penning trap on the outside of the ground electrode end ring) in the entrance/exit region (see Figs. 6.4 and 6.8). The major design element of the new electrode is that it has no end ring. It is manufactured from a single aluminum block and its shape is exactly adjusted to the curvature of the magnetic field lines.

With this electrode design the background rate could be reduced by 5 orders of magnitude to about 30 mHz. This result finally did convict the small Penning trap on the outside of the ground electrode to be the culprit behind the "High-B-Field" background. Notwithstanding this success an underlying question remains: How can such a small Penning trap produce a kHz background?

6.2.2 Penning Trap Induced Background via Photons

At first, simulations showed that positive ions produced in the pre-spectrometer Penning trap move, unaffected by the magnetic field, straight to the negative anti-Penning electrode. Electrons produced there, however, follow the magnetic field lines and are guided to the ground electrode far away from the Penning trap. Therefore, the filling mechanism of the trap could not be due to positive ions (as one would typically expect). The sequence of processes is visualized in Fig. 6.9.

In comparison photons can reach the end ring area of the anti-Penning electrode. From there, electrons, produced by photo-effect are directly guided back to the position of the Penning trap. When accelerated to the anti-Penning electrode potential, these electrons have a small probability to ionize residual gas, thereby producing

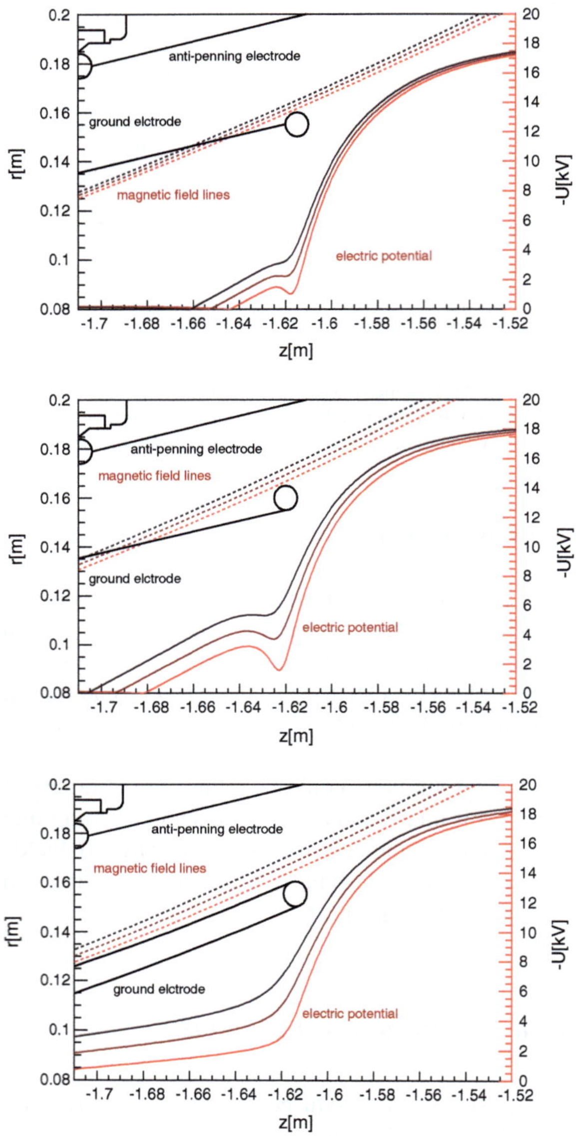

Fig. 6.8 Penning trap at ground electrode in setup I-III. The figure shows three magnetic field lines (*dashed lines*) running just above the end ring of the ground electrode. The *solid lines* represent the electric potential (y-axis on the *left*) along these magnetic field lines. *Top* The electric potential has a minimum along the magnetic field lines of $U_{\text{trap}} = 0.9$ kV, *Middle* The Penning trap is increased to $U_{trap} = 2.5$ kV. *Bottom* In this setup all Penning trap have been eliminated

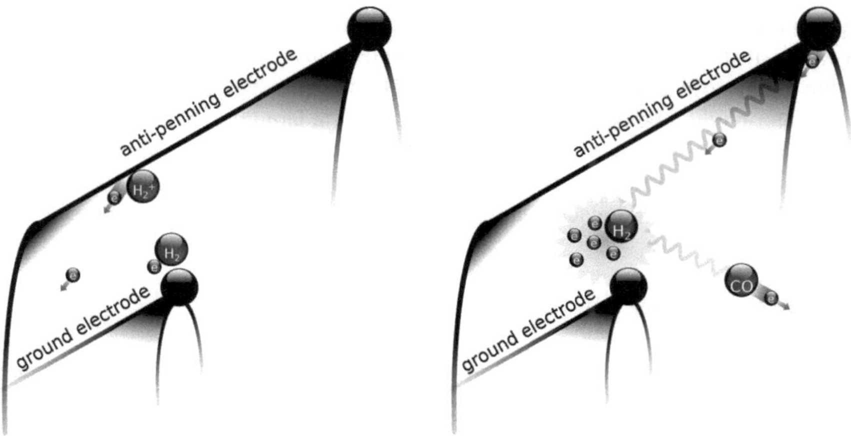

Fig. 6.9 Background production of Penning traps at the pre-spectrometer. *Left* The figure demonstrates that the background production mechanism cannot proceed via positive ions. The ions hit the anti-Penning electrode just above the ground electrode and produce electrons that can never enter the volume of the Penning trap again. Therefore, a positive feedback mechanism is ruled out. *Right* The Penning trap also produces UV-light. Not influenced by electro-magnetic fields, the photons illuminate the whole volume of the pre-spectrometer. Some photons will hit the end of the anti-Penning electrode, thereby produce electrons via photo-effect. These electrons reach the volume of the trap and inevitably fill it up. Photons flying into the sensitive volume of the spectrometer can photo-ionize residual gas producing electrons that can reach the detector

secondary electrons, which subsequently are stored in the trap. This procedure is visualized in Fig. 6.9. With a probability of about 30 % the electron from the cathode (anti-Penning electrode) are backscattered on the ground electrode end ring. In this case the cross section for ionization is increased due to the reduced kinetic energy of the backscattered electron.

The electrons stored in the trap have no direct line of sight with the detector, neither can positive ions move into the spectrometer. However, photons can illuminate the sensitive volume of the spectrometer. There, the photons produce secondary electrons via photo ionization and in this way create the observed background rate.

In the following corroborating facts for this phenomenological model of background production via photon emitting Penning traps are listed.

Photon Production of Trap

In the literature the phenomenon of Penning traps acting as a UV-light source is widely discussed by a number of papers [7–9]. A direct experimental confirmation of this phenomenon occurring actually in the pre-spectrometer setup was provided by following measurement: It is possible to set the e-gun on high potential without the UV-light being switched on. With this setting electrons starting on the surface of the e-gun have enough energy to overcome the retarding potential and reach the detector. However, if the UV-light is switched off, no electrons are expected to be produced on the e-gun surface. In the presence of the "High-B-Field" background an image of the e-gun tip could be observed at the detector, i.e. a large number of electrons was created on the e-gun tip without the UV-light being switched on. This observation gives a very strong hint that the Penning trap is producing photons. Photons are the only particles that could reach the e-gun tip from the inside of the spectrometer [10].

Pressure Dependence

Both the rate of photons produced in the trap and the rate of secondaries created in the volume depend on the pressure in the spectrometer. Consequently, to first order a quadratic dependence of the background rate on the pressure is expected. Measurements revealed exactly this behavior, as shown in Fig. 6.10.

Measured Rate

In setup II a kHz background rate was measured. This large rate raises the question whether such a small Penning trap is capable of producing enough photons to explain this measurement result.

In the following, a simple estimation of the number of photons being created in the Penning trap present in the pre-spectrometer will be shown.

The scattering probability P for UV photons in the residual gas is given by

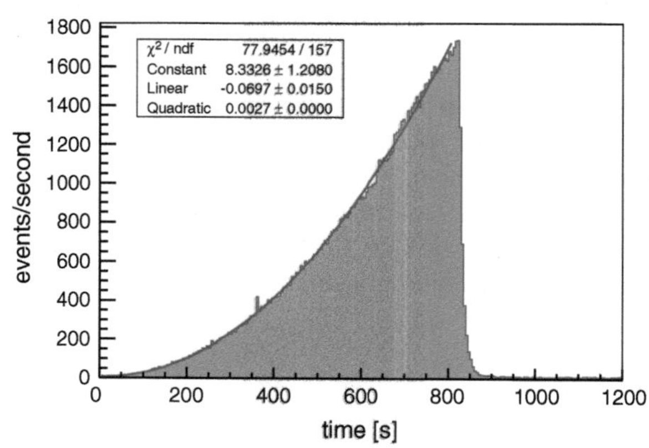

Fig. 6.10 Pressure dependence of background rate. The measurement shows that the background rate rises quadratically with the pressure, as expected

$$P = 1 - e^{-\left(l \cdot \frac{p}{k_B \cdot T} \cdot \sigma\right)} \tag{6.1}$$

$$\approx l \cdot \frac{p}{k_B \cdot T} \cdot \sigma, \quad \text{(for small p)} \tag{6.2}$$

where l denotes the path length of the photon, p the pressure, k_B the Boltzmann constant, T the temperature and σ the cross section. The probability for photo-ionization of hydrogen turns out to be $P_{hydro} \approx 10^{-7}$, with a cross section of $\sigma = 5 \cdot 10^{-18}$ cm^2 (for $E_{photon} = 14$ eV), a path length of $l = 1$ m and a pressure of $p = 10^{-9}$ mbar. Correspondingly, to explain a kHz background rate (i.e. 10^3 electrons/s) a rate of at least 10^{10} photons/s must be produced in the Penning trap.

At first the question of how many electrons can be stored in the pre-spectrometer Penning trap is addressed. The trap has a rather small volume of

$$V_{Penning} = 2\pi r \cdot A \approx 75\,\text{cm}^3, \tag{6.3}$$

where the cross sectional area A of the Penning trap is given by a width of $\Delta z = 1.5$ cm and a height of $\Delta r = 0.5$ cm. For simplicity the electrons are considered to be stored in a torus like volume around the end ring of the ground electrode ($r_{Penning} = 16$ cm and an intrinsic radius $R = 0.5$ cm) as shown in Fig. (6.11). Furthermore, to first order a constant charge distribution inside the trap can be assumed.

To find the number of stored electrons, following question needs to be answered: how many electrons inside the torus create a potential U_{equi} at which the Penning trap is stable? As a first step the electric field $\vec{E}(r)$ for $(r < R)$ created by the charge density ρ in the torus is computed with the help of Gauss's law. For this calculation the origin is chosen to be at $r_{Penning}$, i.e. in the center of the cross section of the torus:

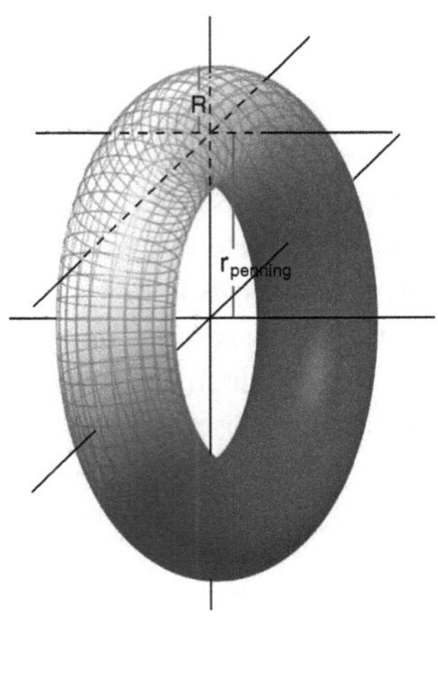

Fig. 6.11 Simplified volume of the Penning trap. To estimate the number of electrons being stored in the Penning trap, a torus like volume of the trap is assumed

$$\int_{\mathcal{O}} \vec{E} dA = \int_{\mathcal{V}} \frac{\rho}{\epsilon_0} \tag{6.4}$$

$$\vec{E}(r) = \frac{\rho}{2\epsilon_0} r\hat{r} \tag{6.5}$$

The electric potential $U(r)$ obeying the boundary condition $U(0) = 0$, is then given by

$$U(r) = \frac{-\rho}{4\epsilon_0} r^2 \tag{6.6}$$

Rearranging Eq. (6.6) provides an expression for the charge density ρ as a function of the potential U. The charge density itself is just the number density of electrons n times the electric charge e. Solving for the number density of electrons we find

$$n = \frac{4U}{r^2} \frac{\epsilon_0}{e} \tag{6.7}$$

From this relation we can deduce the following: one needs an electron density of n in the torus to create an electric potential of U. For the Penning trap in setup II the trap depth is $U_{\text{trap}} = 2.5$ kV and $r = R \approx 5 \cdot 10^{-3}$ m. Assuming an equilibrium potential of $U_{\text{equ}} = \frac{1}{2} U_{\text{trap}}$ an electron number density of

$$n = 1.2 \cdot 10^{10}\,\text{cm}^{-3} \tag{6.8}$$

follows. With the volume given in Eq. (6.3), the number of electrons N in the trap is

$$N = 9.5 \cdot 10^{11}. \tag{6.9}$$

To find the corresponding number of photons being produced by this number of electrons per second, the probability to produce one photon (here: to excite a H_2 gas molecule) is computed. The cross section for excitation of H_2 by a 100 eV electron is $\sigma = 10^{-20}$ cm^2. The path length of a stored electron in 1 s is $l(1s) \approx 10^7$ m, the partial pressure of H_2 is assumed to be $p = 10^{-9}$ mbar and the temperature is $T = 300$ K. Hence, with Eq. (6.1), the probability for an electron of 100 eV to excite a H_2 gas molecule in 1 s is found to be $P \approx 0.63$.

With this results, one can estimate the number of photons produced by $N = 9.5 \cdot 10^{11}$ electrons per second to be $N_{\text{photons}} \approx 6 \cdot 10^{11}$, which is sufficient to explain the observed background rate of about the order of kHz.

6.3 Conclusion

The test measurements at the pre-spectrometer, described in this chapter, could show that even small Penning traps (V $<$ 100 cm^3), located on the outside of the inner electrode system can cause tremendous background rates of up to several kHz. Precise electromagnetic field calculations were necessary to find the Penning trap and to eliminate it. A phenomenological model to explain the background production mechanism was developed and confirmed by measurements.

To avoid Penning trap induced background in the final setup, the electromagnetic design has to ensure that the creation of Penning traps is ruled out. The ground electrode of the main spectrometer has been redesigned in view of the observations at the pre-spectrometer. Furthermore, intensive simulations searching for possible Penning traps in the complete spectrometer volume and detector section have been performed [11].

The elimination of this background source has paved the way to the discovery of the next layer of background. This entirely new background source is smaller in its absolute rate, however, has the potential to seriously limit the neutrino mass sensitivity of KATRIN, as well.

References

1. X. Liu, D.E. Shemansky, Ionization of Molecular Hydrogen. Astrophys. J. **614** (2004)
2. C. Heise, J. Hollandt, R. Kling, M. Kock, M. Kuhne, Radiometric characterization of a Penning discharge in the vacuum ultraviolet. Appl. Opt. **33** (1994)
3. T. Tabata, T. Shirai, Analytic cross sections for collisions of H+, H2+, H3+, H, H2, and H- with hydrogen molecules. At. Data Nucl. Data Tables **76**(1), 1–25 (2000)
4. J.A.R. Samson, Photoionization of atoms and molecules. Phys. Rep. (Sect. C Phys. Lett.) 28(4) (1976)
5. F. Fränkle, Background investigations of the KATRIN pre-spectrometer. PhD thesis (KIT, 2010)
6. F. Habermehl, Electromagnetic measurements with the KATRIN pre-spectrometer. PhD thesis (KIT, 2009)
7. E.S. Warden, H.W. Moos, Penning discharge as a photoelectric EUV spectroscopy source. Appl. Opt. **16**(1902–1904) (1977)
8. S. Bowyer, Continuous emission source covering the 50–300-A band. Appl. Opt. **32** (1993)
9. D.S. Finley, S. Bowyer, F. Paresce, R.F. Malina, Continuous discharge Penning source with emission lines between 50 A and 300 A. Appl. Opt. **18** (1979)
10. F. Fränkle, Connection between e-gun background and high b-field background. Electron. Logbook Entry
11. T. Corona, Penning trap searches in the detector region. KATRIN collaboration meeting, Oct 2011

References

Chapter 7
Background Due to Stored Electrons Following Nuclear Decays

A previously underestimated background source, first being discovered at the pre-spectrometer [1, 2], arises from magnetically trapped high energy electrons. The main source of these electrons are ^{219}Rn and ^{220}Rn α-decay as well as tritium β-decays, in the volume of the KATRIN spectrometers. On the basis of the design values for tritium retention and allowed levels of radioactive impurities of all main spectrometer components, the expected background level will exceed the required bounds. As a consequence of the findings of this work more stringent requirements have to be set on the vacuum system of the spectrometer. Furthermore, both active and passive background reduction mechanisms will have to be applied.

In this chapter the general background production mechanism of electrons from nuclear α- and β-decays will be outlined. Thereafter, a specific radon background model, validated by measurements at the pre-spectrometer, will be presented. Finally, the expected background rate as well as the consequences for the design and the sensitivity of KATRIN will be discussed.

7.1 Model of the Background Production Mechanism

The MAC-E-Filter principle inherently relies on the magnetic field at the entrance and exit region to be much higher compared to the magnetic field in the center of the spectrometer. At the main spectrometer the magnetic field drops by a factor of $2 \cdot 10^4$ from the entrance to the center. However, this setup forms a magnetic bottle for light charged particles.

An electron produced in the center of the pre- or main spectrometer is accelerated towards the ends of the spectrometer thereby moving from a low magnetic field region into a high magnetic field region. Consequently, its longitudinal energy is transformed into transversal energy. Depending on the starting angle and energy of the electron, all its kinetic energy can be transformed into transversal energy before the electron leaves the spectrometer and hence this electron is magnetically trapped. The trajectory of a stored electron is visualized in Fig. 7.1.

S. Mertens, *Background Processes in the Electrostatic Spectrometers of the KATRIN Experiment*, Springer Theses, DOI: 10.1007/978-3-319-01177-6_7,

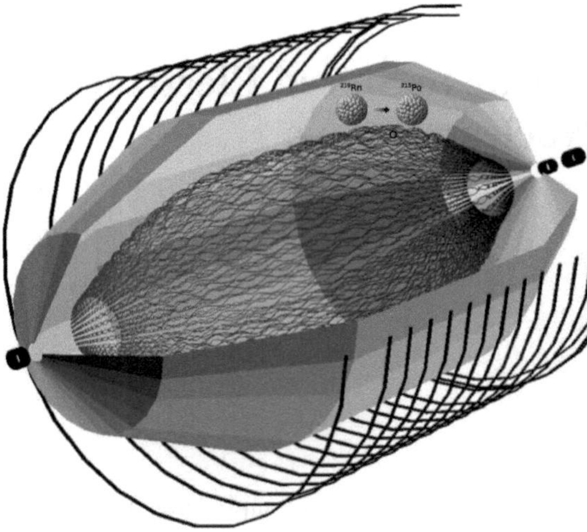

Fig. 7.1 An electron produced via ^{219}Rn α-decay in the volume of the main spectrometer has a probability of over 90 % to be stored due to the magnetic mirror effect. It performes axial, cyclotron and magnetron motion, see also Fig. 5.4. During its storage time it ionized residual gas. The subsequent secondary electron can reach the detector and produce background

Eventually, every stored electron looses energy due to inelastic and elastic scattering on residual gas molecules and synchrotron radiation. If the electron is cooled down to energies of the order of eV it is eventually released from the trap. Due to the low pressure of $p = 10^{-11}$ mbar maintained in the KATRIN main spectrometer, the cooling down time can reach up to about 10 h. During this time the electron produces several hundred low-energy secondary electrons via ionization of residual gas molecules. Eventually, these secondary electrons leave the magnetic trap and reach the detector.

The main source of high-energy primary electrons are nuclear decays occurring in the volume of the spectrometer. Of concern for KATRIN are: ^{219}Rn and ^{220}Rn α-decays, as well as tritium β-decays. The isotope ^{219}Rn arises from the ^{235}U chain, whose decay product ^{231}Pa is contained in small quantities in the NEG getter material (SAES NEG strips St707) used for pumping in the spectrometers. Apart from this dominant source, ^{220}Rn from the ^{232}Th decay chain emanates from material that is used in the interior of the main and pre-spectrometer, such as e.g. glass insulators. The well known isotope ^{222}Rn, which is of specific interest for low-level underground experiments such as Majorana, GERDA or Borexino, is not an issue for KATRIN, since its lifetime is longer than the pumping time, so it has no chance to decay inside the spectrometer volume. Finally, a very tiny fraction of the order of 10^{-14} of the tritium flow rate from the Windowless Gaseous Tritium Source (WGTS) can reach the spectrometer section of KATRIN, and decay there.

Depending on the number density of the decaying atom, and the absolute pressure which governs the storage time, several electrons can be stored simultaneously, which increases the background rate respectively.

7.2 Radon Model to Explain Observations at Pre-spectrometer

After the removal of Penning trap-related background a long-term measurements (50 h) at a pressure of $p = 1 \cdot 10^{-10}$ mbar was performed at the pre-spectrometer with symmetric magnetic field $B = 4.5$ T and retarding potential $U_{ret} = -18$ kV (run 1a of [1]). The measurements showed time periods ($\Delta t = 15$–120 min) of increased rates up to 250 mHz. The pixel distribution during the times of elevated rate featured characteristic ring structures centered at the beam axis of the spectrometer (see Fig. 7.2).

These observations can only be explained by stored high energy electrons. The stored electron performs an $\vec{E} \cdot \vec{B}$ and a $\vec{\nabla}|\vec{B}| \cdot |\vec{B}|$ magnetron drift, superimposed on its axial motion and the much faster cyclotron motion (see Fig. 5.4). During this drift it ionizes residual gas molecules and thereby produces low-energy secondary

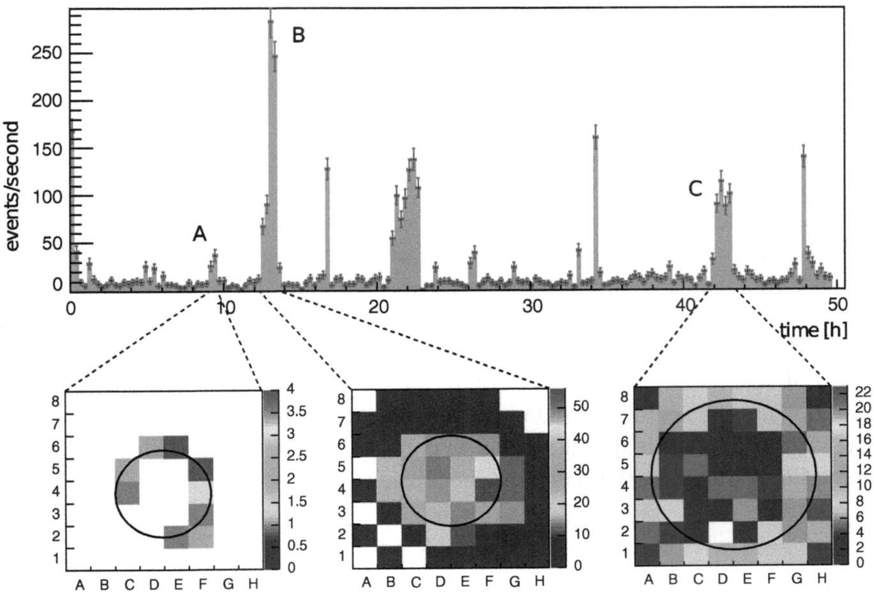

Fig. 7.2 During a 50 h background run [1] specific periods of an elevated background rate were observed. The rate can rise up to 250 mHz, and the interval of increased rate can last up to about several hours. During the periods of elevated rate, ring structures of variable radii are observed at the pixel detector

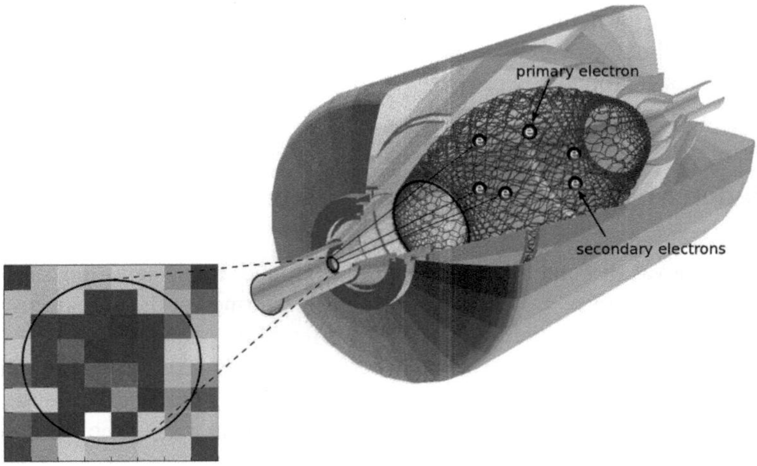

Fig. 7.3 The observed ring like structure at the pixel detector during times of elevated rate (see Fig. 7.2 C), can be explained by stored electrons. A trapped electron performs a azimuthal magnetron motion superimposed on its axial and cyclotron motion. Secondary electrons are produced via ionization of residual gas, are low energetic and can therefore leave the magnetic trap by following the magnetic field line they are created on. Consequently, they arrive on a ring like distribution at the detector

electrons. These are not stored and fly towards the detector, following the magnetic field lines. As visualized in Fig. 7.3, the secondary electrons arrive on a ring like distribution on the pixel detector.
The number of secondary electrons reaching the detector during a period of enhanced rate (ring events) sheds light on the energy of the primary stored electron. In the simplest approximation one can assume that the primary electron looses about $\langle E_{\text{loss}} \rangle = 50$ eV per ionization. The number of secondaries during a ring event varies between 50 and 5,000, indicating primary electron energies of $E_{\text{primary}} = 2.5 - 250$ keV.

The remaining open question, at this point, is how fairly high-energy electrons in the keV range can be created inside the volume of the pre-spectrometer. The best candidate for such an electron source are the α-decays of ^{219}Rn and ^{220}Rn. In the following the sources of radon and the electron emission via α-decay will be discussed in detail. Finally, the radon hypothesis will be validated by comparing Monte Carlo simulation with measurement results.

7.2.1 Emanation of Radon

In the UHV regime of the spectrometers, only the noble gas isotopes ^{220}Rn, ^{222}Rn and ^{219}Rn are of relevance for KATRIN background:

- ^{219}Rn from the ^{235}U decay chain (uranium-actinium chain): half life $t_{1/2} = 3.96$ s

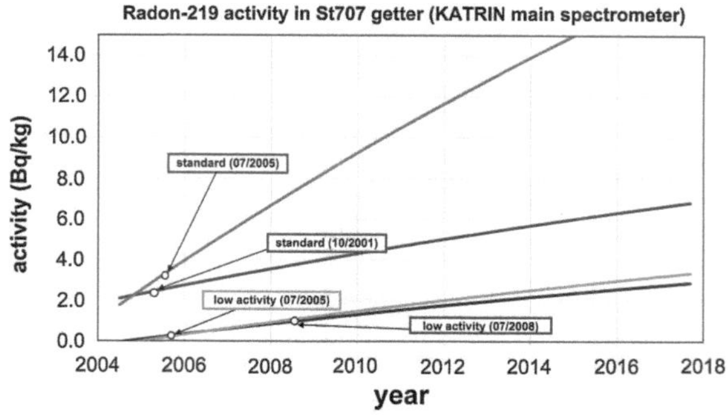

Fig. 7.4 ^{219}Rn activity of different batches of *SAES St707* getter alloy. An excess of ^{231}Pa ($t_{1/2} = 32\,760$ a) decays into ^{227}Ac ($t_{1/2} = 21.8$ a), which decays into short-lived isotopes, including ^{219}Rn. During the test measurements with the pre-spectrometer the getter alloy labeled 'standard (10/2001)' was used. For the final getter pumps of the pre-spectrometer and main spectrometer a special 'low-activity' alloy is used, which radon activity is reduced by a factor of two

- ^{220}Rn from the ^{232}Th decay chain (thorium chain): half life $t_{1/2} = 55.6$ s
- ^{222}Rn from the ^{238}U decay chain (uranium-radium chain): half life $t_{1/2} = 3.82$ d

These three decay chains (see Fig. 7.5) are the most common chains observed on the earth, all end in a different stable isotope of lead.

Despite careful radio assaying and use of low-activity components, the structural elements of the spectrometers contain small traces of elements from these decay series, which subsequently can decay into radon. One such material is zirconium, which constitutes 70 % of the getter alloy *SAES St707* used in getter pumps. Due to the porous getter surface, as visual in Fig. 7.6, the noble gas ^{219}Rn can easily emanate.

Figure 7.4 shows the measured activity of different batches of the getter alloy as a function of calendar year. Even low-activity batches contain elements of the actinium series, which produce ^{219}Rn. Since the decay series is not in secular equilibrium the radon production rate is still increasing over time.

7.2.2 Electron Creation by Radon Decay

The α-decay of a radon isotope is in itself no direct source of background, as the α-particle and the daughter ion are too heavy to follow the magnetic field lines. It is the emission of electrons following the α-decay that is of concern here. Figure 7.7 visualizes all possible electron creation mechanisms, which will be disscussed in detail in the following. Figure 4.2 shows the electron energy spectra of ^{219}Rn and ^{220}Rn α-decay.

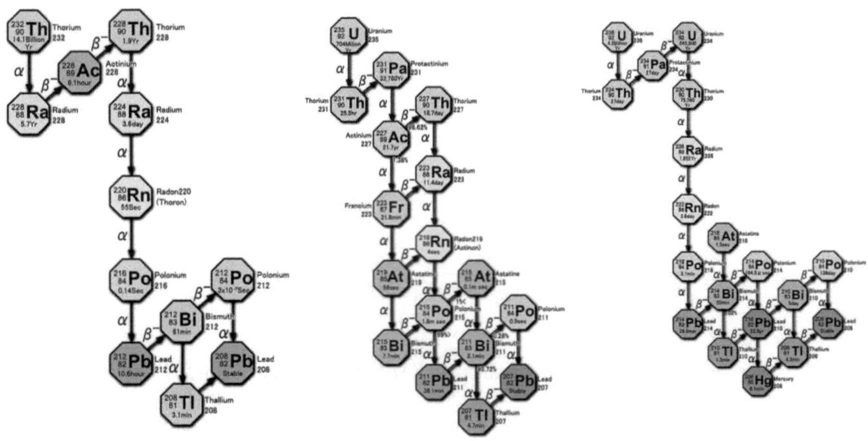

Fig. 7.5 ^{220}Rn occurs in the decay chain of thorium (*left*). ^{219}Rn occurs in the uranium-actinium decay chain (*middle*), ^{222}Rn occurs in the uranium-radium chain (*right*) [3–5]

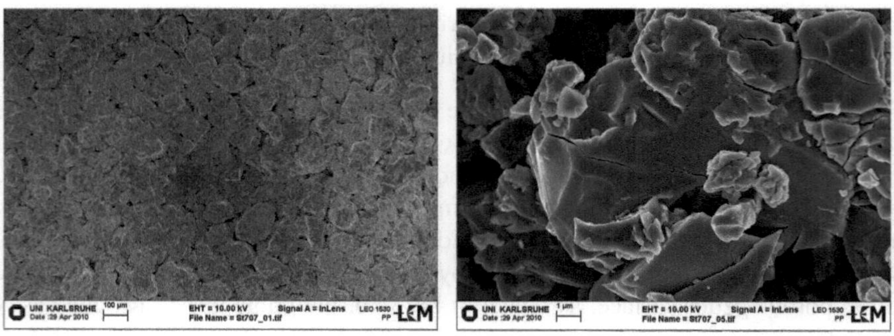

Fig. 7.6 These pictures of the getter material were obtained by an electron microscopy at KIT's LEM laboratory. *Left* Grains of typical size of 100 μm are visible. *Right* Zoom into one of the grains

Conversion Electrons

The radon isotopes decay via α-decay into polonium daughter atoms, ^{215}Po in case of ^{219}Rn, and ^{216}Po in case of ^{220}Rn. ^{215}Po is an odd nucleus, and, consequently, features a large variety of excited nuclear states. In contrast, ^{216}Po, being an even-even nucleus, is not as easily excited. It has only one excited state that can be populated via the ^{220}Rn α-decay. As a consequence, ^{219}Rn has a rather high probability to decay into one of the excited nuclear states of ^{216}Po. Whereas, ^{220}Rn has only a small chance to decay into the excited nuclear state of ^{216}Po.

The excited polonium state decays into the ground state within a period of picoseconds. If the wave function of a shell electron is non-vanishing at the position of the

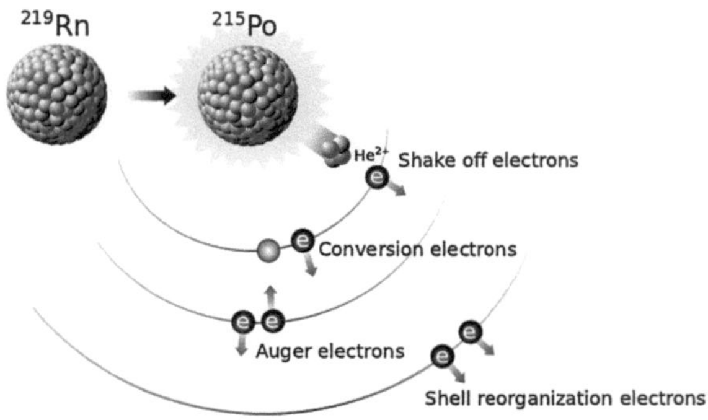

Fig. 7.7 The figure visualizes the different electron production mechanisms in ^{219}Rn α-decay. Details of the mechanisms are explained in the text

nucleus, it can be emitted in the de-excitation process, instead of radiating gammas or X-rays. These electrons are called conversion electrons.

Shake-Off and Shell Reorganization Electrons

The α-particle is ejected from the nucleus with a typical velocity of $v_\alpha = 10^7$ m/s. The velocities v_e of the K, L and M shell electrons are comparably large to this speed ($\frac{v_\alpha}{v_e} \approx 0.1$). Therefore, for the inner shell electrons the change of the charge of the nucleus can be considered adiabatically. The probability of electron emission due to a fast reduction of the central field can be neglected. Instead a sizable emission probability arises from a direct collision of the α-particle with the electron. In this case, the ejected electrons are called shake-off electrons.

Electrons of the outer shells have velocities significantly lower than the α-particle velocity, ($\frac{v_\alpha}{v_e} \approx 10$). As a result the collision probability is low, however, the change of the charge of the nucleus is perceived sudden, i.e. non-adiabatic, by the outer electrons [6]. This sudden change of the central field leads to an emission of some outer shell electrons, the so called shell reorganization electrons. Measurements show that the charge distribution of the recoiling polonium peaks at zero [7]. Consequently, since the α-particle takes away 2 protons, in most cases 2 shell reorganization electrons are emitted.

Since the electron shell configuration is almost identical for ^{219}Rn and ^{220}Rn, the probabilities for shake-off electron emission and the energies of these electrons can be assumed to be identical.

Auger Electrons

Conversion electrons as well as shake-off electrons leave a vacancy in the electron shell. These vacancies are filled in turn by electrons from outer shells thereby emitting photons of the freed-up energy. However, it is possible that instead of a photon an

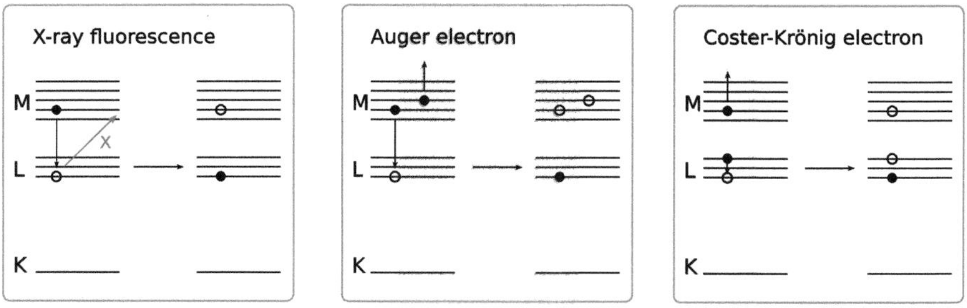

Fig. 7.8 When a vacancy in the electron shell is filled by an electron of an outer shell the freed-up energy can result in the emission of a photon or an Auger electron. In case of Auger electron production two vacancies are left behind. If the electron filling the vacancy is of the same shell as the vacancy, the emitted particles are called Coster-Krönig electrons

electron, the so called Auger electron, is emitted (see Fig. 7.8). Auger electrons often occur in cascades of relaxations: The electron filling the vacancy as well as the Auger electrons leave a new vacancy, which is filled again etc. The energies of Auger electrons and the relative probabilities of Auger emission are assumed to be identical for ^{219}Rn and ^{220}Rn, owing to the almost identical shell structures.

7.2.3 Validation of Radon Model

In the following several arguments will be listed which clearly underline that the background is indeed due to stored electrons originating from ^{219}Rn and ^{220}Rn decays in the volume of the pre-spectrometer.

Comparison of Simulated and Measured Rates

To understand the observed time periods with an elevated background rate in detail a Monte Carlo simulation of the ^{219}Rn and ^{220}Rn decay processes in the pre-spectrometer was performed. For this purpose a radon generator including the production of all types of primary electrons (as described in Sect. 4.3.1) was implemented into KASSIOPEIA.

A quantitative comparison of simulation and measurement was achieved by parameterizing the experimental data in the following way: each ring event is described by the duration of the elevated rate and the number of electrons hitting the detector during this time. Accordingly, the parameters: duration of event Δt and number of detector hits N_{det} are compared to the simulation. For this comparison experimental data of all radon background measurements [1] was combined.

The simulation ensemble consisted of 7000 ^{219}Rn and ^{220}Rn decays in the pre-spectrometer operated at a pressure of $1 \cdot 10^{-10}$ mbar. For each decay, the number

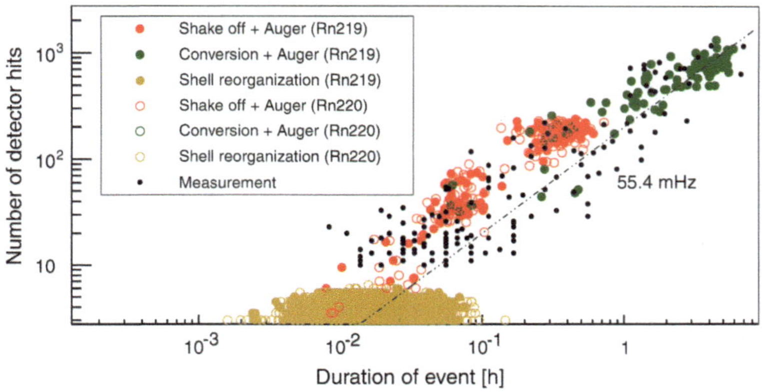

Fig. 7.9 The plot shows all measured ring events (*black dots*) parametrized by their duration and the number of electrons reaching the detector during the time of elevated rate. The data [1] were taken at different pressure levels in the pre-spectrometer. As this changes the duration of the events, the measured event durations were corrected to a pressure of $1 \cdot 10^{-10}$ mbar. The colored dots represent the simulation of 7,000 radon decays in the pre-spectrometer with KASSIOPEIA. The detailed simulation settings are given in Table E.1

of electrons reaching the detector in the energy region of interest as well as the total length of an event were determined.

Figure 7.9 shows that there is very good agreement between simulation and measurement. Two basic conclusions can be drawn from this comparison:

- The times of elevated background rate seen in the measurement arise from conversion and shake-off electrons and their subsequent Auger electrons.
- Shell reorganization electrons can explain the underlying constant background of a few mHz in between the time periods of elevated rate.

The multiplicity of shell reorganization electrons is too small to be counted as ring events. However, in about 94 % of all radon decays shell reorganization electrons are produced, which makes them a good candidate to explain the background rate in between the time periods of enhanced rate. In the following an estimation to reinforce this statement is given:

Considering a background measurement in the standard configuration (run 1a of [1]), the average background rate during time of non-elevated rate is (3.2 ± 0.3) mHz. On the other hand, on average (8.2 ± 2.9) ring events occur per day. This number determines the actual number of ^{219}Rn and ^{220}Rn decays, since from the simulation it is known that in about 6 % of all decays a ring event occurs. Hence, with a probability for shell reorganization electron emission of 94 %, (128.5 ± 45.4) shell reorganization events would be expected in this measurement.

As in each shell reorganization event on average 2.6 electrons reach the detector in the energy region of interest. A corresponding background rate of (3.9 ± 1.4) mHz would be expected from shell reorganization electron emission. From this estimation

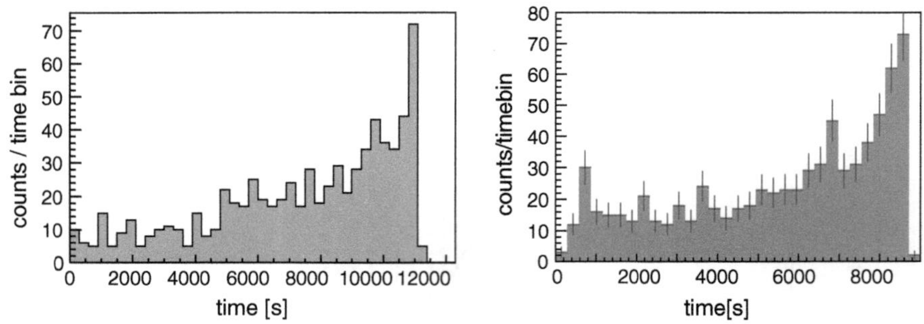

Fig. 7.10 The figures show the event rate at the detector as a function of time. As described in the text, the rate increases towards the end of the event. This is expected since the cross section is maximal for about 100 eV, see Fig. 7.11, which the primary electron only reaches when it is almost fully cooled down. *Left* Simulation with KASSIOPEIA. *Right* Measurement [1]

one can conclude that the shell reorganization electrons are an excellent candidate to explain the entire underlying background rate of (3.2 ± 0.3) mHz.

This excellent agreement is one of the main outcomes of this work: in the pre-spectrometer the only sources of background stem from Penning traps (see Chap. 6) and the α-decay of the two isotopes ^{219}Rn and ^{220}Rn.

Time Structure of Increased Rate

An interesting feature of the time periods with enhanced rate is their intrinsic time structure. Typically, the rate rises towards the end of the event. This can be nicely explained considering the energy dependence of the electron cross section. The primary stored electron starts off with an energy of several keV and slowly cools down. Shortly before it cools below the minimal energy to be stored ($E_{\mathrm{min}} = 70$ eV) it reaches the maximal cross section for ionization at about 100 eV (see Fig. 7.11). Simulations follow this characteristic time behavior as shown in Fig. 7.10. This observation further reinforces the hypothesis that the observed background is caused by stored electrons in the keV regime.

Ring Radius Distribution

As illustrated in Fig. 7.3 the stored electrons lead to ring-like pixel distributions at the detector. Due to the huge mean free path of the noble gas atoms ^{219}Rn and ^{220}Rn, one can assume a homogeneous distribution of α-decays in the spectrometer. Consequently, the distribution of the ring radius is expected to be linearily increasing with radius. Both simulation and measurement confirm this prediction, see Fig. 7.12. With this analysis it is proven that the source of primary electrons is homogeneously

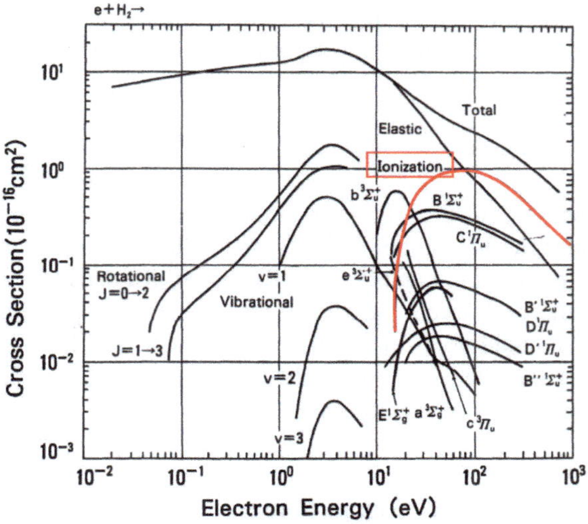

Fig. 7.11 For the background considerations, only the ionization cross section, marked in *red* is most relevant [8]. The figure shows that the ionization cross section is maximal at about 100 eV

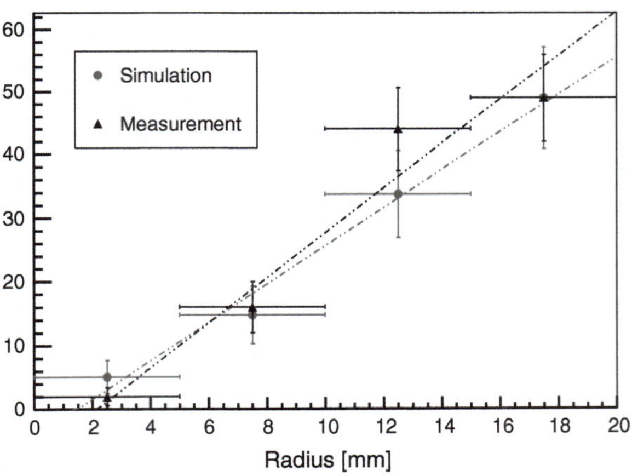

Fig. 7.12 The histogram shows the radii distribution of 118 measured ring events (*red circles*). This is compared to the simulated ring radius distribution with KASSIOPEIA. The simulation is normalized to the number of events with a ring radius $15 < r < 20$ mm. A linear fit on simulated and measured data gives a slope of $m_{\mathrm{sim}} = 3.00 \pm 0.87$ and $m_{\mathrm{meas}} = 3.52 \pm 0.92$, respectively . The linear increase of the radius confirms the hypothesis of homogeneously distributed radon isotopes in the spectrometer

distributed in the volume of the pre-spectrometer. This in turn substantiates the assumption that the primary electrons origin from ^{219}Rn and ^{220}Rn α-decay.

Number of Radon Decays

By comparing measurements at different pressure levels and gas compositions it is possible to disentangle the number of ^{219}Rn and ^{220}Rn in the pre-spectrometer [1]. Another way to find out the number of ^{219}Rn and ^{220}Rn decays is to compare the measurement results at different pressure regimes with corresponding simulations.

By fitting the simulation results to the measurement, and letting the number of decays be a free parameter, one finds values that are in good agreement with the results obtained with the first method, see Table 7.1.

Pressure Dependence of Background Rate

The background rate measured at the detector is determined by the number of secondary (and tertiary) electrons that a primary electron will produce and the corresponding primary's storage time. The number of secondary electrons is largely determined by the primary energy available for ionization. The storage time also depends on the primary energy: The larger its initial energy, the longer it takes for the primary electron to cool down. The over-all storage time also scales with the pressure: The higher the pressure the shorter the ionization time and consequently the storage times. However, the number of secondaries is independent of the pressure.

Consequently, when increasing the pressure, shorter time intervals of increased rates with correspondingly higher background rates are expected. Measurements with an artificially increased pressure showed exactly this expected characteristics (see [1]). These measurements can be considered a further experimental proof of the hypothesis that stored high-energy electrons cause the observed background.

Table 7.1 Comparison of simulated and measured radon activity in pre-spectrometer

Source	Activity [mBq] from simulation with KASSIOPEIA, this work	Activity [mBq] deduced from measurement only [1]
^{220}Rn	54.6 ± 7.4	33 ± 9
^{219}Rn from walls	1.3 ± 1.1	2.4 ± 2.0
^{219}Rn from getter	4.55 ± 2.1	7.5 ± 1.8

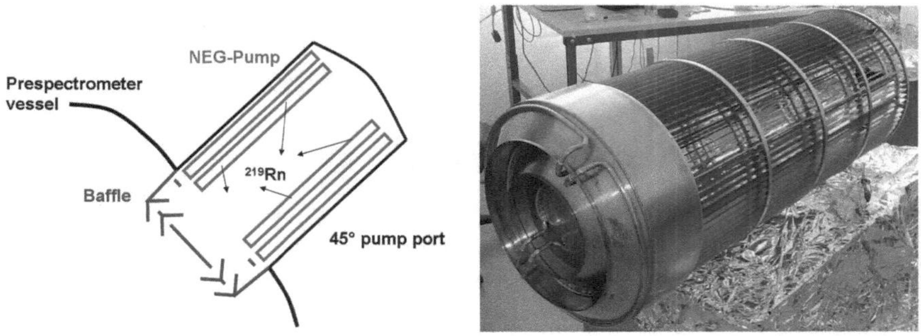

Fig. 7.13 *Left* schematic view of installation of baffle between getter pump and prespectometer vessel. The ^{219}Rn emanating from the getter has no line of sight to the pre-spectrometer volume and is freezing to the nitrogen cooled baffle surface. *Right* Photograph of the baffle installed at the getter pump of the pre-spectrometer

No Getter and Cooled Baffle

As mentioned above, the getter material is known to be a source of ^{219}Rn. An obvious choice was thus to perform a measurement without the getter. A distinct reduction was observed [9]. When removing other parts of the spectrometer under suspicion to emanate ^{220}Rn, such as glass windows and thermo elements, the occurrence of ring events could be further reduced [9].

Most importantly, it could be shown that installing a nitrogen cooled baffle in front of the getter pump yields the same result as the measurements without getter [9]. The baffle is designed in a way that there is no direct line of sight between the pump and the spectrometer volume. Accordingly, all ^{219}Rn emanating from the getter material hits at least once the surface of the baffle. By cooling the baffle to 77 K the ^{219}Rn atoms stick to its surface for a time period longer than several ^{219}Rn lifetimes and cannot enter the volume of the spectrometer. A schematic view and a photograph of the baffle are shown in Fig. 7.13.

Comparison of Radon Model to Independent Measurement

In [7] the charge distribution of ^{216}Po in the α-decay of ^{220}Rn was measured. Since the α-particle takes away two positive charges, the ^{216}Po charge is zero if in addition two electrons are emitted. Accordingly, one can translate the charge distribution into a distribution of the number of primary electrons in the decay. This distribution can be simulated by KASSIOPEIA. In Fig. 7.14 the measured distribution from the literature is compared to the simulated one.

The good agreement between both distributions most notably underlines that the rather complex electron emission processes following a radon α-decay have been fully understood and implemented successfully.

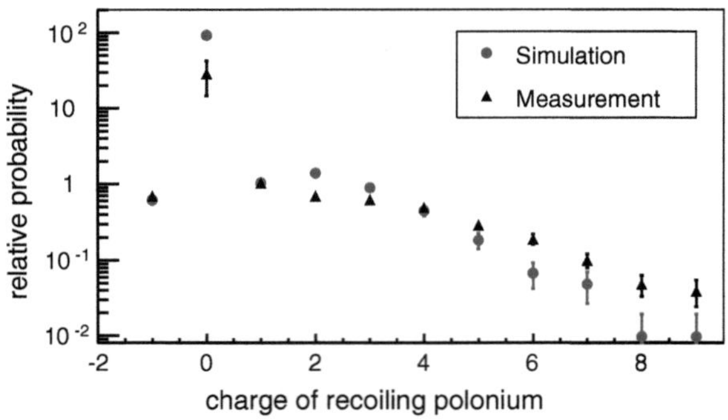

Fig. 7.14 The result of the full KASSIOPEIA simulation is shown in *red circles*, the corresponding measurement of [7] in *black dots*. The agreement is excellent for most charge recoils. The deviation at zero might be explained by the fact that, as stated in [7] the value at a ^{216}Po charge of zero has only been estimated. Furthermore, a comparison to the pre-spectrometer measurement, as reported in Sect. 7.2.3, confirms the ratio of neutral ^{216}Po to charged ions: A ^{216}Po charge of zero corresponds almost exclusively to shell reorganization events, where only two low-energy electrons are ejected. With the ratio of shell reorganization events (94 %) to other events (6 %: conversion electrons, shake-off and subsequent Auger electrons) from the simulation, the entire background rate can be explained

7.3 Expected Background Rate at the Main Spectrometer

In the previous Sect. 7.2, it could be shown that the background model due to ^{219}Rn and ^{220}Rn α-decays is in very good agreement with the observations at the pre-spectrometer.

In the main spectrometer an even larger amount of ^{219}Rn and ^{220}Rn α-decays can be expected. Instead of only 90 m getter strips, a total of 3 km getter strips will be installed at the main spectrometer. To reduce the number of ^{219}Rn in the spectrometer, a passive shielding, consisting of a liquid nitrogen cooled baffle, as it was tested already at the pre-spectrometer (see above), will have to be used.

At the main spectrometer, yet another source of background of a rather similar type will occur: This is tritium, which is neither pumped in the differential pumping section (DPS) nor in the cryogenic pumping section (CPS) and that therefore can enter the spectrometers and decay there. The subsequent β-decay electrons are stored in the main spectrometer as the electrons emitted in radon α-decay. Tritium β-decay in the sensitive volume of the spectrometer, will therefore lead to a comparable background characteristics as discussed above for the α-decays of ^{219}Rn and ^{220}Rn. Even with the design value for tritium reduction factor of 10^{14}, the reference background limits are exceeded. Hence, new constraints on the vacuum system have to be considered and the pre-spectrometer will have to be redesigned as additional tritium pump.

In the main spectrometer the magnetic field in the center is much weaker than in the pre-spectrometer. This fact changes the magnetic storage conditions compared to the pre-spectrometer. On the one hand this leads to an increased storage probability for low-energy electrons, on the other hand the storage probability of high-energy electrons is decreased.

The reference pressure at the main spectrometer of $p^{\mathrm{MS}} = 1 \cdot 10^{-11}$ mbar as compared to the pressure in the pre-spectrometer of about $p^{\mathrm{PS}} = 5 \cdot 10^{-10}$ mbar leads to a fundamentally different background situation. The storage time of the trapped electrons as a result of the lower pressure is increased by almost an order of magnitude. This fact changes the background characteristics drastically, since in this situation the time periods of enhanced rate can reach up to 12 h and are thus very likely to overlap. The background rate by single decays is reduced (since the number of produced secondaries stays the same), however, the time periods of elevated rate can no longer be taken out from data analysis.

The goal of this section is to estimate the total expected background rate from tritium and radon decay in the main spectrometer. In a first step, the storage conditions of the main spectrometer are investigated in detail (Sect. 7.3.1). Thereafter, in Sect. 7.3.2 the actual background rate expected by single tritium and radon decays is simulated. An estimation of the tritium and radon decay rates in the main spectrometer volume is presented in Sect. 7.3.3. In Sect. 7.3.4 the results are combined and the absolute expected background rate is shown for different vacuum scenarios.

7.3.1 Stored Electrons in the Main Spectrometer

In the following the storage probabilities of electrons in the main and pre-spectrometer will be compared in detail. In particular, the energy dependence of the trapping probabilities will be investigated. Analogously, the dependence of the number of secondaries on the primary electron energy is subject of this section.

Detailed Comparison of Storage Conditions in the Main and Pre-spectrometer

As outlined above, in the main spectrometer the storage conditions are different from the pre-spectrometer. The low magnetic field in the analyzing plane of the main spectrometer of $B_{\mathrm{ana}}^{\mathrm{MS}} = 3 \cdot 10^{-4}$ T ($B_{\mathrm{ana}}^{\mathrm{PS}} = 156 \cdot 10^{-4}$ T), leads the fact that already electrons of energies $E_{\mathrm{trap}}^{\mathrm{MS}} > 1$ eV ($E_{\mathrm{trap}}^{PS} > 70$ eV) have a non-vanishing trapping probability.

On the other hand, the low magnetic field leads to a disruption of the storage conditions for high-energy electrons: If the transversal starting energy of the electron exceeds a critical value of $E_{\mathrm{break}} = 180$ keV its cyclotron radius is larger than the radius of the main spectrometer, and therefore the electron directly hits the wall on a timescale $t_{\mathrm{break}} < 10^{-7}$ s.

In addition, the motion of high-energy electrons in low magnetic fields is non-adiabatic. The transformation of $E_{\perp}$ into $E_{\parallel}$ and vice versa is no longer proportional to the change of the magnetic field, i.e. the angle of the momentum to the magnetic

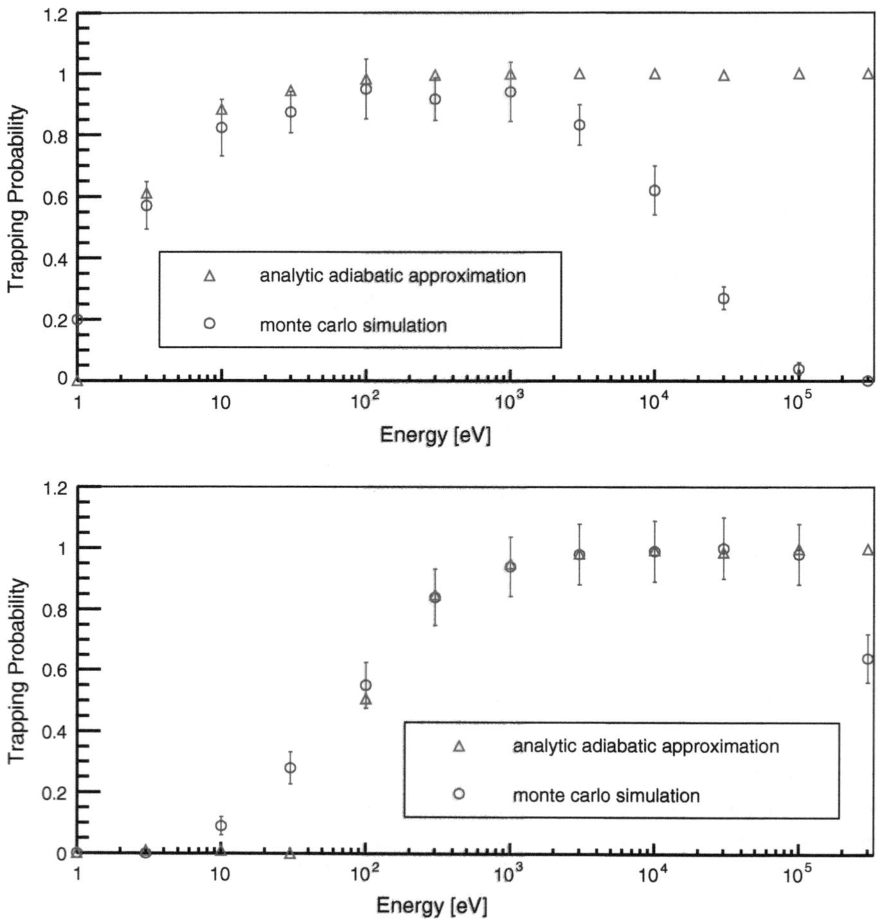

Fig. 7.15 In this figure the result of a full Monte Carlo simulation with KASSIOPEIA which takes into account non-adiabatic effects is compared to an adiabatic analytic calculation of the trapping probability (see Appendix G.1). The simulation settings are given in Table E.2. *Top* At the main spectrometer the trapping probability increases fast for low energies since the probability of the electron to start with large transversal energy increases with its total starting energy. At large energies the trapping probability decreases due to non-adiabatic effects. *Bottom* At the pre-spectrometer the storage probability rises slower, reflecting the inferior energy resolution of the pre-spectrometer. However, at high energies the storage probability is higher due to the stronger magnetic field in the center, resulting in a better adiabaticity as compared to the main spectrometer

field line changes "randomly" [10]. Therefore, non-adiabatic motion allows the electron to escape the magnetic mirror trap. The escape time at very large energy is of the order of $t_{\text{break}} < 10$ ms. In Fig. 7.15 the calculated storage probabilities of electrons in the main and pre-spectrometer are compared.

Number of Secondaries as a Function of Primary Energy

A primary high-energy stored electron eventually cools down via ionization of residual gas. The secondary electrons it produces are mainly low energetic and leave the spectrometer on a rather short time scale of minutes. Considering only energy loss due to ionization of gas molecules, the number of secondaries produced by a primary electron of fixed energy is approximately given by

$$N_{\mathrm{e}}(E_{\mathrm{primary}}) \approx \frac{E_{\mathrm{primary}}}{\langle E_{\mathrm{loss}} \rangle}, \tag{7.1}$$

where E_{loss} describes the average energy the primary looses per ionization. However, the following corrections need to be taken into account:

- secondary electrons themselves can be stored again thereby producing more tertiary electrons,
- the storage probability is decreased for high energy-electrons, due to non-adiabatic effects and large cyclotron radii, see Fig. 7.15
- stored electrons additionally loose energy via synchrotron radiation, elastic scattering and excitation of residual gas molecules

Figure 7.16 shows the number of secondary electrons as a function of starting energy. At very low energies no storage is possible. At very large energies non-adiabatic effects break the storage condition and therefore shorten the storage time, during which the production of secondary electrons would take place. The correlation of the number of secondaries to the primary starting energy allows, in principle, a spectroscopy of stored particles.

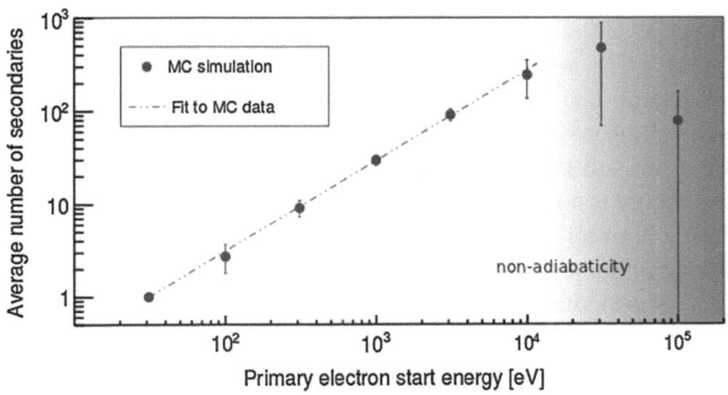

Fig. 7.16 The simulation settings for this figure are given in Table E.3. A clear linear correlation between number of secondaries and primary energy emerges until the non-adiabatic regime

7.3.2 Background Rate Due to Single Tritium β- and Radon α-Decays

Before the absolute background rate at the main spectrometer can be estimated, the rate created by single nuclear tritium β- and radon α-decays needs to be understood. The background characteristic created by both decays is rather similar, however, basic differences can be pointed out: Several electrons are emitted by radon α-decays whereas a tritium β-decay only a single electron is produced. Moreover, the energy spectra of ^{219}Rn, ^{220}Rn and tritium decays are very different: Tritium has a continuous β-sectrum reaching up to 18.6 keV, whereas ^{219}Rn, ^{220}Rn feature distinct mono-energetic energy lines reaching up to more than 100 keV. In the following it will be discussed how these differences influence the resulting background rate.

Radon

As described in Sect. 7.3.1 electrons of energies $E > 100$ keV only weakly contribute to the background rate, due to their large cyclotron radii and non-adiabatic effects. Based on this fact, and considering the electron energy spectrum of ^{219}Rn decays (see Fig. 4.2), it is evident, that the high energy conversion electrons are not the dominant background source at the main spectrometer.

In the case of ^{220}Rn conversion electrons of $E_{\text{conversion}} = 450\,(534)$ keV are produced with a probability of $3 \cdot 10^{-5}$ $(6 \cdot 10^{-6})$. Consequently, owing to their small branching ratios and high energies their background contribution can be completely neglected.

The main background contribution, hence, arises from shake-off electrons and Auger electrons subsequent to conversion and shake-off processes. Shake-off and Auger electrons lie in the few keV energy range, consequently, rather long storage times can be expected. The probability for the creation of Auger electrons in ^{219}Rn decay is $P_{\text{Rn219}} = 6\,\%$, whereas for ^{220}Rn one obtains $P_{\text{Rn220}} = 3.8\,\%$.

If no conversion and no shake-off electron is produced on average two shell reorganization electrons are created in the "neutralisation" of the daughter ion. These electrons lie in the sub-keV energy region. Hence, their storage probability is high, however, the energy is not sufficient to create a large number of secondary electrons.

In total 97.0 % of all ^{219}Rn and ^{220}Rn α-decays lead to background production (i.e. at least one of the primary electrons is stored and creates secondary electrons).

Figure 7.17 shows the number of secondary electrons per radon decay together with the overall duration of the enhanced background level as a scatter plot. The plot allows to discriminate the different mechanisms leading to electron emission. However, the discrimination is not as distinct as compared to the pre-spectrometer (see Fig. 7.9). In Appendix G.2 more details about the characteristics of the background due to ^{219}Rn and ^{220}Rn α-decays can be found.

These results are in particular important for an understanding of background processes in the main spectrometer test measurements in 2012.

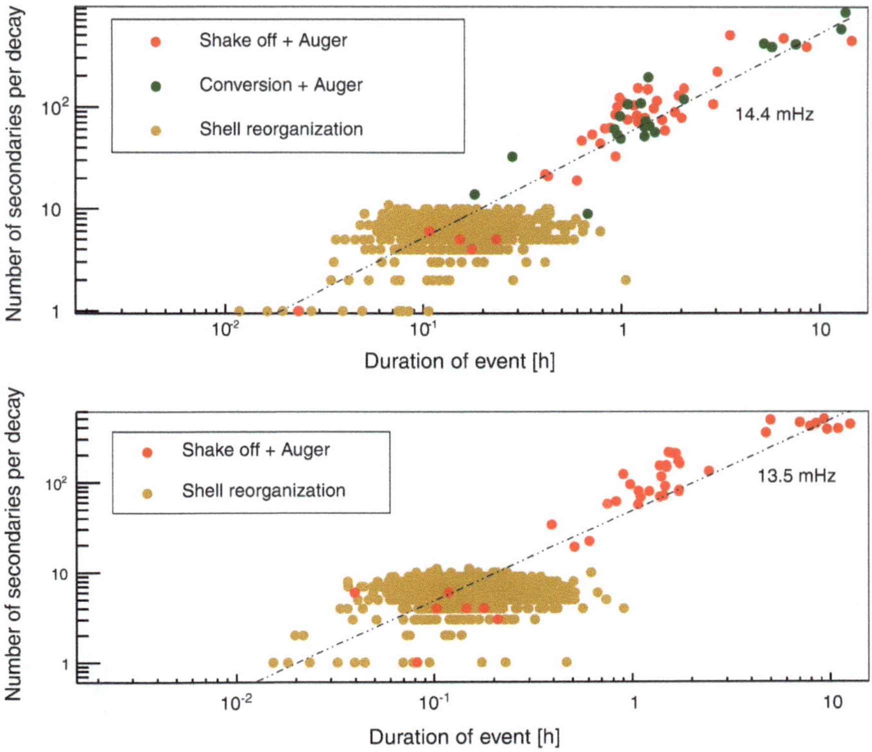

Fig. 7.17 *Top* Conversion and shake-off processes of ^{219}Rn α-decay and in particular the subsequent Auger electron emission will lead to enhanced background rates of more than 10 h. Conversion and shake-off processes are not distinguishable as in the case of the pre-spectrometer, since the conversion electrons themselves are not stored, but only their subsequent Auger electrons. The frequent shell reorganization electrons are stored for up to one hour in the main spectrometer and contribute significantly to the overall background. *Bottom* In the case of ^{220}Rn α-decay the contribution from conversion processes can be neglected. The simulation settings for this figure are given in Table E.4

Tritium

The storage probability of single electrons of less than 18.6 keV from tritium β-decay is rather high, as expected in this energy regime (see Fig. 7.15). However, since in contrast to a radon decay, only one electron is created, only 57 % of all tritium decays lead to background production. Figure 7.18 shows the number of secondaries and the duration of the enhanced rate as a function of the primary start energy for tritium β-decay compared to ^{219}Rn α-decay.

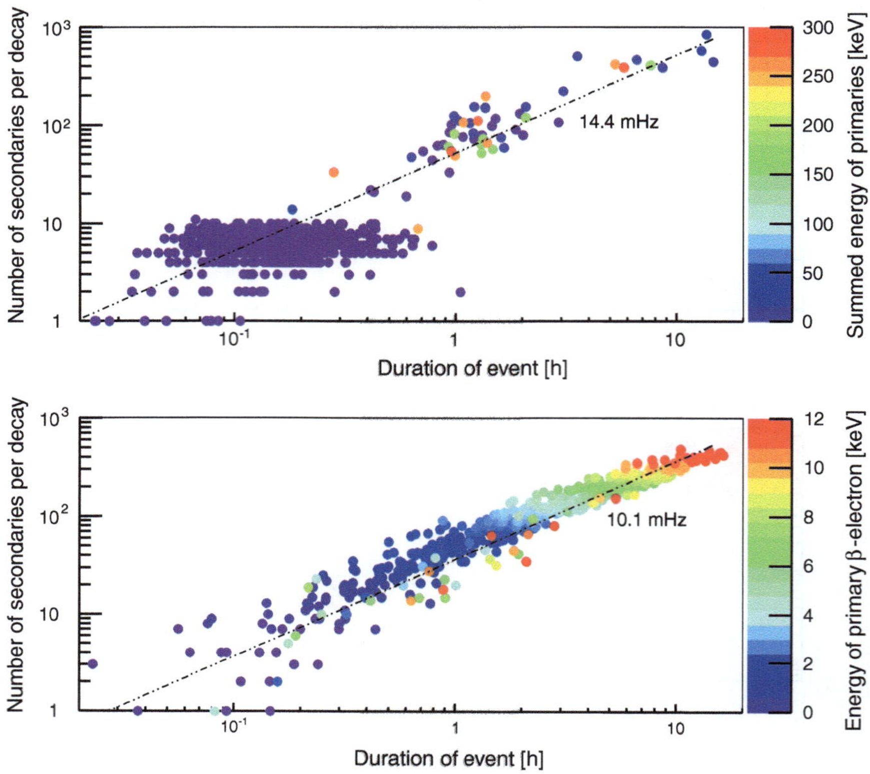

Fig. 7.18 *Top* The color coding indicates the sum of the primary electron energies of a ^{219}Rn α-decay. *Bottom* In case of tritium β-decay a clear correlation between the primary energy and the event duration as well as the number of secondaries is visible. This spectrum is of crucial importance when defining the maximal tritium flow rate into the main spectrometer. The simulation settings for this figure are given in Table E.4

7.3.3 Estimation of Nuclear Tritium and Radon Decay Rates in the Main Spectrometer

Based on measurements at the pre-spectrometer [1], the radon emanation rates from the getter material and other sources can reliably be estimated and extrapolated to the main spectrometer. Assuming a constant radon and tritium density in the spectrometer, one can derive the expected decay rates in the spectrometer volume.

As it will be shown in the following, an increasing amount of getter material in the pre-spectrometer decreases the contribution of tritium to the background, on the other hand, the radon decay rate is increased. As a countermeasure, liquid nitrogen cooled cryo-baffles could be installed in front of the three pump ports of the main spectrometer, causing however a decrease of hydrogen (tritium) pumping speed.

In the following considerations, the nuclear decay rates will be given for different amounts of getter material in the pre-spectrometer and the implications of a scenario with and without a cryo-baffle will be outlined [11]. At the end of this section, four exemplary scenarios will be singled out.

Radon Emanation Rates

The measurements at the pre-spectrometer revealed the dominant source of ^{219}Rn emanation to be the getter material. Further sources such as vacuum gauges or a temperature sensor were identified. After removing the getter pump and these components, still a small number of radon-like events was observed.

For the extrapolation to the main spectrometer a worst case scenario is assumed, in which the radon (which does not explicitly originate from the getter material) originates from the stainless steel walls. Thus the radon emanation is scaling with the spectrometer vessel surface (surface area pre-spectrometer: 25 m^2, main spectrometer: 690 m^2).

In the main spectrometer a low-activity batch of the NEG getter material with a factor of 2 lower activity will be used. The estimation of the resulting radon emanation rate from this getter material is based on Fig. 7.4. The radon emanation rates of the standard getter material used in the pre-spectrometer, as well as the special low-activity getter and the stainless steel walls are given in Table 7.2.

Radon Decay Rate

Based on the emanation rates given in Table 7.2 and the characteristics of the vacuum system of the pre- and main spectrometer, as described in detail in Appendix F the number of radon atoms and corresponding decay rates can be determined. The number of radon atoms $N_{\mathrm{Rn}}^{\mathrm{MS}}$ inside the main spectrometer can be calculated by solving the following differential equation:

Table 7.2 Radon emanation rate from different sources

		Emanation rate	
Isotope	Source	Absolute	Normalized
^{219}Rn	Standard getter	$(7.5 \pm 1.8) \cdot 10^{-3}$ Rn/s	$(8 \pm 2) \cdot 10^{-5}$ Rn/(s · m)
^{219}Rn	Low-activity getter	$(3.8 \pm 0.9) \cdot 10^{-3}$ Rn/s	$(4 \pm 1) \cdot 10^{-5}$ Rn/(s · m)
^{219}Rn	Stainless steel wall	$(1 \pm 1) \cdot 10^{-3}$ Rn/s	$(4 \pm 4) \cdot 10^{-5}$ Rn/(s · m^2)
^{220}Rn	Stainless steel wall	$(3 \pm 2) \cdot 10^{-3}$ Rn/s	$(12 \pm 8) \cdot 10^{-5}$ Rn/(s · m^2)

Depending on the source of radon, the activity has been normalized to the pre-spectrometer surface area (25 m^2) and getter length (90 m), respectively. The low-activity getter, which will replace the standard getter in the final setup could not be measured in the pre-spectrometer. Its two times lower activity was estimated using Fig. 7.4. The ^{219}Rn rate from the wall is based on a single event observed in the pre-spectrometer during a five day background measurement

$$\frac{dN_{\mathrm{Rn}}^{\mathrm{MS}}}{dt} = -\lambda_{\mathrm{Rn}} \cdot N_{\mathrm{Rn}}^{\mathrm{MS}} - \frac{N_{\mathrm{Rn}}^{\mathrm{MS}}}{V_{\mathrm{MS}}} \cdot S_{\mathrm{eff}}^{\mathrm{MS}}(\mathrm{Rn}) + E_{\mathrm{Rn}}^{\mathrm{PS}\rightarrow\mathrm{MS}} + E_{\mathrm{Rn}}^{\mathrm{MS}}. \tag{7.2}$$

The four main contributions to the differential equation are:

- the total decay rate of radon: $A_{\mathrm{Rn}}^{\mathrm{MS}} = \lambda_{\mathrm{Rn}} \cdot N_{\mathrm{Rn}}^{\mathrm{MS}}$,
- the amount of radon pumped out by the TMPs and the baffle: $\frac{N_{\mathrm{Rn}}^{\mathrm{MS}}}{V_{\mathrm{MS}}} \cdot S_{\mathrm{eff}}^{\mathrm{MS}}(\mathrm{Rn})$,
- the radon influx from the pre-spectrometer $E_{\mathrm{Rn}}^{\mathrm{PS}\rightarrow\mathrm{MS}}$ and
- the emanation rate $E_{\mathrm{Rn}}^{\mathrm{MS}}$ of radon from various sources in the main spectrometer (see Table. 7.2).

The effective pumping speeds of the two TMPS in the pre-spectrometer are given in Table F.1. The baffles can capture radon which is present in the volume of the main spectrometer and therefore work as additional radon pump with an effective pumping speed of

$$S_{\mathrm{baffle}} = \alpha \cdot \frac{\overline{c}}{4} \cdot A, \tag{7.3}$$

where α denotes the sticking coefficient of the nitrogen cooled baffle, $\overline{c}$ the average velocity of the gas and A the surface area of the baffle.

The radon flowing back from the main spectrometer to the pre-spectrometer can be neglected, since the radon density in the main spectrometer is much lower than in the pre-spectrometer.

The number of radon atoms moving from the pre-spectrometer into the main spectrometer through the beam-line valve depends on their number density in the pre-spectrometer and on the conductance of the beam-line valve. With the standard 2 TMP configuration of the pre-spectrometer only 3.6 % of the ^{219}Rn atoms and 13.3 % of the ^{220}Rn atoms emanating in the pre-spectrometer can reach the main spectrometer.

When the system is in equilibrium, the radon density remains constant, and one find the number of radon isotopes $N_{\mathrm{Rn}}^{\mathrm{MS}}$ and the decay rate $A_{\mathrm{Rn}}^{\mathrm{MS}}$ to be

$$\frac{dN_{\mathrm{Rn}}^{\mathrm{MS}}}{dt} = 0 \tag{7.4}$$

$$\rightarrow N_{\mathrm{Rn}}^{\mathrm{MS}} = (E_{\mathrm{Rn}}^{\mathrm{PS}\rightarrow\mathrm{MS}} + E_{\mathrm{Rn}}^{\mathrm{MS}}) \cdot \frac{V_{\mathrm{MS}}}{\lambda_{\mathrm{Rn}} \cdot V_{\mathrm{MS}} + S_{\mathrm{eff}}^{\mathrm{MS}}(\mathrm{Rn})} \tag{7.5}$$

$$\rightarrow A_{\mathrm{Rn}}^{\mathrm{MS}} = \lambda_{\mathrm{Rn}} \cdot N_{\mathrm{Rn}}^{\mathrm{MS}} \tag{7.6}$$

The results for $E_{\mathrm{Rn}}^{\mathrm{MS}}$, $N_{\mathrm{Rn}}^{\mathrm{MS}}$ and $A_{\mathrm{Rn}}^{\mathrm{MS}}$ are listed in Table 7.3 for different getter configurations and sources.

Table 7.3 Radon decays in the main spectrometer

Isotope	Source	Emanation [Rn/s]	No. of atoms	Activity [mBq]
		no baffles		
^{219}Rn	180 m getter (PS)	$(2.6 \pm 0.7) \cdot 10^{-4}$	$(1.5 \pm 0.4) \cdot 10^{-3}$	0.26 ± 0.06
^{219}Rn	180 m + 250 m getter (PS)	$(6.4 \pm 1.6) \cdot 10^{-4}$	$(3.6 \pm 0.9) \cdot 10^{-3}$	0.63 ± 0.16
^{219}Rn	3000 m getter (MS)	$(1200 \pm 300) \cdot 10^{-4}$	$(675 \pm 169) \cdot 10^{-3}$	118 ± 30
^{219}Rn	MS wall	$(276 \pm 276) \cdot 10^{-4}$	$(155 \pm 155) \cdot 10^{-3}$	27.2 ± 27.2
^{220}Rn	MS wall	$(832 \pm 552) \cdot 10^{-4}$	$(5428 \pm 3601) \cdot 10^{-3}$	67.8 ± 45.0
		with baffles		
^{219}Rn	180 m getter (PS)	$(2.7 \pm 0.7) \cdot 10^{-4}$	$(3.0 \pm 0.7) \cdot 10^{-4}$	0.05 ± 0.01
^{219}Rn	180 m + 250 m getter (PS)	$(6.4 \pm 1.6) \cdot 10^{-4}$	$(7.1 \pm 1.8) \cdot 10^{-4}$	0.12 ± 0.03
^{219}Rn	3000 m getter (MS)	$(0 \pm 0) \cdot 10^{-4}$	$(0 \pm 0) \cdot 10^{-4}$	0 ± 0
^{219}Rn	MS wall	$(276 \pm 276) \cdot 10^{-4}$	$(305 \pm 305) \cdot 10^{-4}$	5.3 ± 5.3
^{220}Rn	MS wall	$(832 \pm 552) \cdot 10^{-4}$	$(1123 \pm 745) \cdot 10^{-4}$	1.4 ± 0.9

Radon emanation rate, number of radon atoms and radon activity in the main spectrometer for different getter lengths and with and without baffles in front of the getter pumps in the main spectrometer. For the baffle a sticking coefficient of 0.8 was assumed. In the pre-spectrometer 180 m getter can be installed in the pump port were as additional getter is installed in spectrometer tank itself (see Fig. 7.21)

Tritium Decay Rate

The maximum allowed tritium flow from the WGTS into the pre-spectrometer as given in [12] is approximately $Q_{\mathrm{T}}^{\mathrm{PS}} \approx 10^{-14}$ mbar · l/s $= 2.5 \cdot 10^5$ molecules/s. Due to the large number of adsorption/desorption processes in the transport section the gas flow will be a mixture of HT and T_2 molecules.

The real tritium flow rate into the pre-spectrometer strongly depends on the performance of the differential (DPS) and cryogenic (CPS) pumping section (see Sect. 3.1.2). With first measurements at the DPS2-F one could extrapolate a reduction factor of the tritium flow of $5 \cdot 10^4$, which is only a factor of two less than the design value [13]. Test measurements of the tritium flow suppression factor of the CPS with the TRAP experiment reached a value of $3 \cdot 10^7$ [14]. Compared to the final CPS cryostat, the TRAP cryo-pump had a 10 times smaller cryo-sorption surface and a tube with only one bend instead of four as in the CPS. Therefore, the suppression factor of the CPS is expected to be better than the value of 10^7 in the design report [12].

The decay rate of T_2 is $\lambda_{\mathrm{T}_2} = 3.58 \cdot 10^{-9}\ \mathrm{s}^{-1}$ and $\lambda_{\mathrm{HT}} = 1.79 \cdot 10^{-9}\ \mathrm{s}^{-1}$ for HT, respectively. The number of tritium molecules $N_{\mathrm{HT}}^{\mathrm{MS}}$ (and analogously for $N_{\mathrm{T}_2}^{\mathrm{MS}}$) inside the main spectrometer, following a similar ansatz as for radon, is obtained by

$$\frac{dN_{\mathrm{HT}}^{\mathrm{MS}}}{dt} = -\lambda_{\mathrm{HT}} \cdot N_{\mathrm{HT}}^{\mathrm{MS}} - \frac{N_{\mathrm{HT}}^{\mathrm{MS}}}{V_{\mathrm{MS}}} \cdot S_{\mathrm{eff}}^{\mathrm{MS}}(\mathrm{HT}) + Q_{\mathrm{HT}}^{\mathrm{PS}\rightarrow\mathrm{MS}}. \tag{7.7}$$

The contributions here are:

- the total decay rate of tritium: $A_{\mathrm{HT}}^{\mathrm{MS}} = \lambda_{\mathrm{HT}} \cdot N_{\mathrm{HT}}^{\mathrm{MS}}$,

- the amount of tritium pumped out by the TMPs and the NEG getter pump: $\frac{N_{\mathrm{HT}}^{\mathrm{MS}}}{V_{\mathrm{MS}}} \cdot S_{\mathrm{eff}}^{\mathrm{MS}}(\mathrm{HT})$ and
- the tritium influx from the pre-spectrometer $Q_{\mathrm{HT}}^{\mathrm{PS}\rightarrow\mathrm{MS}}$.

The effective pumping speeds for tritium can be found in Table F1. Again, the ("backward") gas flow from the main to the pre-spectrometer can be neglected. Furthermore, due to the long lifetime of tritium as compared to the pump-down time, also the tritium decays $\lambda_{\mathrm{HT}} \cdot N_{\mathrm{HT}}^{\mathrm{MS}}$ inside the spectrometer can be neglected. Under equilibrium conditions one finds:

$$\frac{dN_{\mathrm{HT}}^{\mathrm{MS}}}{dt} = 0 \tag{7.8}$$

$$\rightarrow N_{\mathrm{HT}}^{\mathrm{MS}} = Q_{\mathrm{HT}}^{\mathrm{PS}\rightarrow\mathrm{MS}} \cdot \frac{V_{\mathrm{MS}}}{S_{\mathrm{eff}}^{\mathrm{MS}}(\mathrm{HT})} \tag{7.9}$$

$$\rightarrow A_{\mathrm{HT}}^{\mathrm{MS}} = \lambda_{\mathrm{T}} \cdot N_{\mathrm{HT}}^{\mathrm{MS}} \tag{7.10}$$

The results for $Q_{\mathrm{HT}}^{\mathrm{PS}\rightarrow\mathrm{MS}}$, $N_{\mathrm{HT}}^{\mathrm{MS}}$ and $A_{\mathrm{HT}}^{\mathrm{MS}}$ are given in Table 7.4.

Table 7.4 Tritium decays in the main spectrometer

Source	PS getter pump	Flux (1/s)	No. of molecules	Activity (mBq)
MS vacuum system with TMPs only, no getter pumps				
HT	0 m	66198	$6.64 \cdot 10^6$	11.9
HT	180 m	3127	313738	0.56
HT	180 m + 250 m	1292	129602	0.23
T_2	0 m	59264	$6.12 \cdot 10^6$	21.9
T_2	180 m	3110	321068	1.15
T_2	180 m + 250 m	1289	133059	0.48
MS vacuum system with TMPs, 3000 m getter strips and no baffles				
HT	0 m	66198	114092	0.20
HT	180 m	3127	5390	0.01
HT	180 m + 250 m	1292	2226	0.004
T_2	0 m	59264	124689	0.45
T_2	180 m	3110	6543	0.023
T_2	180 m + 250 m	1289	2712	0.01
MS vacuum system with TMPs, 3000 m getter strips and baffles				
HT	0 m	66198	248324	0.44
HT	180 m	3127	11731	0.02
HT	180 m + 250 m	1292	4846	0.009
T_2	0 m	59264	270353	0.97
T_2	180 m	3110	14187	0.05
T_2	180 m + 250 m	1289	5879	0.02

The values for T_2 and HT fluxes from the pre-spectrometer into the main spectrometer, the number of molecules and the decay rates in the main spectrometer are given for different configurations of the pre-spectrometer and main spectrometer vacuum systems. The values are to be understood as conservative design values and are therefore given without an error estimation

Four Exemplary Scenarios

In the following, four intuitive, exemplary vacuum scenarios will be singled out.

In comparison to the 90 m getter in use at the pre-spectrometer, a length of 3,000 m getter is used at the main spectrometer, a fact which drastically increases background due to radon decays. As a first countermeasure one might consider using less getter material (→ scenario 1), however, when reducing the amount of getter material one increases the number of tritium atoms decaying in the main spectrometer.

It turns out that the optimal getter length with regard to background is achieved with 180 m getter in the pre-spectrometer pump ports and 250 m additional getter material inside the pre-spectrometer volume, as shown in Fig. 7.20. Figure 7.21 shows a technical drawing of a potential mounting structure for the getter between the vessel and the inner electrode system. In total 180 m low-activity getter are installed in the two pump ports of the pre-spectrometer, 250 m low-activity getter in the volume of the pre-spectrometer and 3,000 m low-activity getter in the main spectrometer (→ scenario 2). In this scenario, where liquid nitrogen cooled baffles are not implemented, the background arising from radon decay exceeds the limits.

Consequently, a liquid nitrogen cooled baffle needs to be installed as a passive barrier at the main spectrometer pump ports (→ scenario 3) (see Fig. 7.19).

The new reference configuration, which is actually being installed at the main spectrometer, is implemented at first without the additional 250 m getter in the pre-spectrometer (→ scenario 4). Since the CPS tritium flow reduction rate is expected to be better than 10^7, the additional getter material in the pre-spectrometer might not be necessary. However, the inner electrode system of the pre-spectrometer is constructed in a way so that additional getter material can easily be installed at the position shown in Fig. 7.21.

- **Szenario 1:** No getter material, to completely avoid background from NEG correlated radon decay

Fig. 7.19 *Left* Photo from the front side. Face-on, there is no line of sight to the getter pumps installed in the pump ports. *Right* The baffle, as seen from above, shows the cylindrical crate holding the NEG getter strips, as well as the cryogenic feed and return lines to cool the baffle to LN2 temperatures

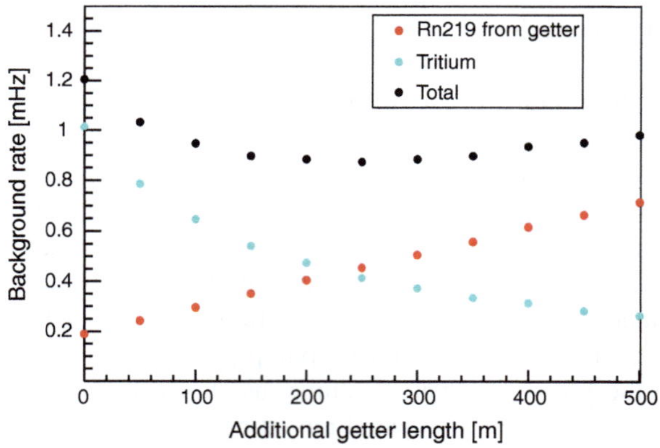

Fig. 7.20 As the plot shows, the radon background contribution increases with increasing getter material whereas the tritium contribution decreases. The optimum is found at 250 m additional getter (180 m getter in pump ports) in the pre-spectrometer volume. The pre-spectrometer will be able to hold up to 1,000 m getter, see Fig. 7.21

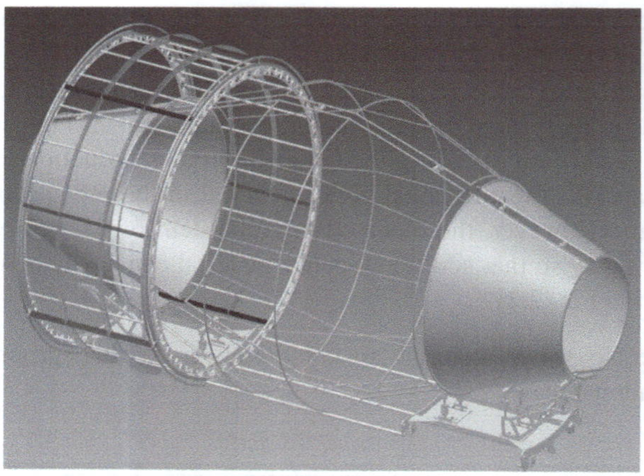

Fig. 7.21 In the figure the inner electrode system of the pre-spectrometer is shown (full cones and wire electrode). The cylindrical holding structure for additional getter is mounted around a full cone of the electrode system. It can hold about 500 m getter strips. A second holding structure, mirrored at the analyzing plane, is foreseen to install another 500 m getter strips, totaling the getter amount to 180 + 500 + 500. At first, no additional getter will be installed, since the tritium reduction by the CPS is expected to be better than the design value

- **Szenario 2:** 250 + 180 m Getter in the pre-spectrometer and 3,000 m getter in the main spectrometer to reduce background arising from tritium decay
- **Szenario 3:** 250 + 180 m Getter in the pre-spectrometer, 3,000 m getter in the main spectrometer and nitrogen cooled baffles installed at the main spectrometer pump ports to simultaneously reduce background arising from tritium and radon decay.

Table 7.5 Expected number of nuclear decays in different UHV scenarios

Scenario	Activity [mBq]			
	Tritium (T2)	$^{219}Rn_{NEG}$	$^{219}Rn_{Wall}$	$^{220}Rn_{Wall}$
1	21.9	0	27.19±27.16	67.85±45.02
2	0.01	118.72±29.52	27.20±27.16	67.85±45.02
3	0.021	0.12 ± 0.03	5.34 ± 5.34	1.40 ± 0.93
4	0.05	0.051 ± 0.013	5.34 ± 5.34	1.40 ± 0.93

The scenarios are described in detail in the main text. Scenario (1–3) represent three extreme cases while scenario (4) is panned to be realized at the start-up of the KATRIN test measurement

- **Szenario 4:** 180 m Getter in the pre-spectrometer, 3,000 m getter in the main spectrometer and nitrogen cooled baffles installed at the main spectrometer pump ports. This is the scenario which will be realized at the start-up of the spectrometer measurements.

Table 7.5 shows the corresponding number of nuclear decays in the main spectrometer in the four different scenarios.

7.3.4 Estimation of Total Background Rate

When combining the results obtained in Sects. 7.3.2 and 7.3.3 one can estimate the total expected background rate for all scenarios. To calculate the number of background events from the number of nuclear decays two factors have to be taken into account:

- The sensitive volume of the main spectrometer covers only 70 % of the total volume (ϵ_V^{MS})
- Only 40 % of all secondary electrons being produced in the main spectrometer will propagate towards the detector, the remaining 60 % exit towards the source side (ϵ_B^{MS}). This asymmetry is due to the non-symmetric magnetic field configuration with the maximum magnetic field reached at the pinch magnet.

The average number of background events $\langle N_{\mathrm{B}} \rangle$ in a time interval T larger than the storage time $T > T_{\mathrm{storage}}$ is given by

$$\left\langle N_{\mathrm{B}}^{\mathrm{MS}} \right\rangle = \epsilon_V^{\mathrm{MS}} \cdot \epsilon_B^{\mathrm{MS}} \sum_i \left\langle N_{\mathrm{d}_i}^{\mathrm{MS}} \right\rangle \left\langle N_{\mathrm{e}_i}^{\mathrm{MS}} \right\rangle, \tag{7.11}$$

where i denotes tritium, ^{219}Rn and ^{220}Rn. The quantity $\langle N_{\mathrm{d}_i} \rangle$ denotes the average number of the decays of the corresponding nucleus in a time interval T. As the partial pressure of tritium and radon is approximately constant, the number of nuclear decays can assumed to be Poisson distributed. $\langle N_{\mathrm{e}_i} \rangle$ denotes the average number of electrons

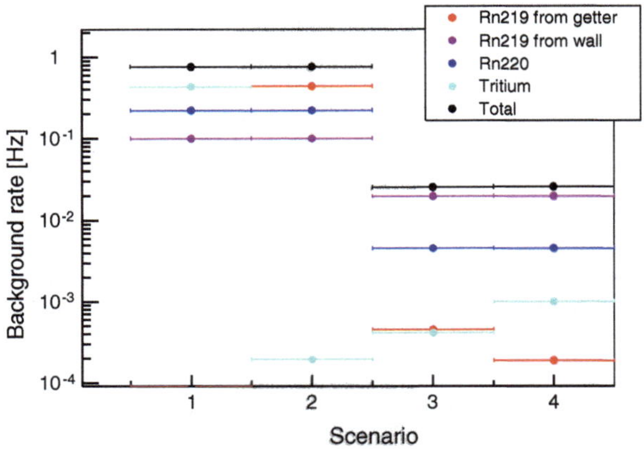

Fig. 7.22 The plot shows the expected background rates for four different UHV scenarios described in Sect. 7.3.3

produced within one event. The distribution of the number of secondary electrons is deduced by the corresponding Monte Carlo simulation as described in Sect. 7.3.2.

Figure 7.22 shows the expected background rate in the four scenarios described in Sect. 7.3.3. Scenario 1 is dominated by the background created by tritium decays. It exceeds with 3.5 Hz the upper bound by more than two orders of magnitude. Scenario 2 still exceeds the background limit by almost 2 orders of magnitude and is dominated by ^{219}Rn from the the getter material. With scenario 3 the background is minimized to 30 mHz. In scenario 4 with no additional getter in the pre-spectrometer the expected background rate is slightly increased.

7.4 Impact of the Background on the Neutrino Mass Sensitivity of KATRIN

In the previous section it could be shown that by installing 250 m additional getter strips in the pre-spectrometer, as well as making use of nitrogen cooled baffle blocking ^{219}Rn atoms emanating from the getter material, a final best case scenario with a background level of about 30 mHz can be achieved.

However, even in this best case configuration the background arising from stored electrons features a special characteristic: The fluctuation of the rate are rather high. As a small number of individual nuclear decays results in a large number of secondary electrons, the count rate of secondary electrons at the detector cannot be expected to be Poisson distributed. The background events are strongly correlated. Therefore, the relative variance of the rate is largely determined by the variance of the significantly smaller number of nuclear decays.

To calculate the neutrino mass sensitivity for this disturbing scenario, a comprehensive statistical study has been performed. Thereby, the integrated spectrum, consisting of signal and background count rates for 41 retarding potentials has been simulated. By fitting the theoretical spectrum to the simulated data the squared neutrino mass m_ν^2 can be extracted. Repeating this procedure N times, a distribution of m_ν^2 is achieved. Its width $\sigma(m_\nu^2)$ gives access to the neutrino mass sensitivity at 90 % confidence level:

$$m_\nu(90\ \%\ \text{C.L.}) = \sqrt{1.64 \cdot \sigma(m_\nu^2)}. \tag{7.12}$$

In a first step it could be shown that large background fluctuations in general would lead to drastic decrease of neutrino mass sensitivity. Assuming for instance a Gaussian distributed background of 30 mHz with a standard deviation of 5 % the neutrino mass sensitivity would be only 0.5 eV at 90 % C.L., i.e. a sensitivity loss of more than a factor of 2 (see Fig. 3.12), as compared to the goal of KATRIN.

To investigate the impact of the realistic background fluctuations arising from stored electrons a detailed background model describing the background as a function of time over the full 3 years measurement time of KATRIN was developed. The model is based on the simulations described in Sect. 7.3.2.

In a typical measurement schedule the integral tritium β-spectrum will be measured at 41 different retarding potentials. The measurement times at each potential are of course fine-tuned to achieve the best neutrino mass sensitivity. In three years of measurement time, there will be many scans over all 41 measurement points. For the detailed background model, the number of secondary electrons during each measurement interval $t_{\mathrm{U}_i}^k$ is simulated (see Fig. 7.23), and superimposed on the signal electrons. The total measurement time at U_i is given by

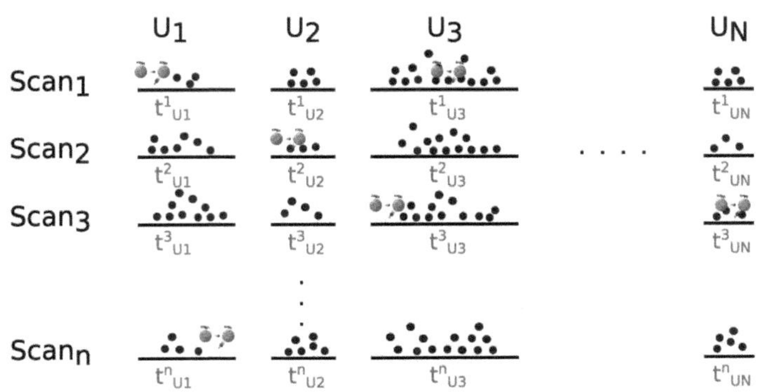

Fig. 7.23 The total measurement time of KATRIN is about 3 years. Within the 3 years n scans of N different retarding potentials will be performed. For the background model the number of secondaries in the different time bins $t_{U_i}^k$ was simulated. First the number of decays in a time bin is determined, then all subsequent secondaries (*red dots*) are distributed into the current and following time bins. The number of secondaries and the storage times are taken from Monte Carlo simulations described in Sect. 7.3.2

$$t_{U_i} = \sum_{k=1}^{n} t_{U_i}^k, \tag{7.13}$$

where k runs over all scans. The time of a scan k is given by

$$t_{scan_k} = \sum_{i=1}^{N} t_{U_i}^k, \tag{7.14}$$

where i runs over all measurement points. A typical scanning time is 3 h. In this case $t_{U_i}^k \approx 5$ min, and the number of scans is approximately 8,000.

Figure 7.24 shows the statistical neutrino mass sensitivity at 90 % C.L. as a function of the overall background rate. Here a Poisson distributed background is compared to the realistic non-Poissonian background model with $t_{scan} = 3$ h and reference pressure $p = 10^{-11}$ mbar. A decrease of sensitivity of about 30 % (at 10 mHz) is observed. The sensitivity at 30 mHz realistic background rate is decreased by about 60 % as compared to the design value.

The simulations in this work have revealed that the fluctuations of the background strongly depend on the time of a measurement interval $t_{U_i}^k$. In other words it is of major importance how fast the different potentials are scanned through. The longer the parameter t_{scan}, the larger the background fluctuations within a measurement interval and the larger the loss of neutrino mass sensitivity. The reason behind this dependence is that the fluctuations within a measurement interval are suppressed if

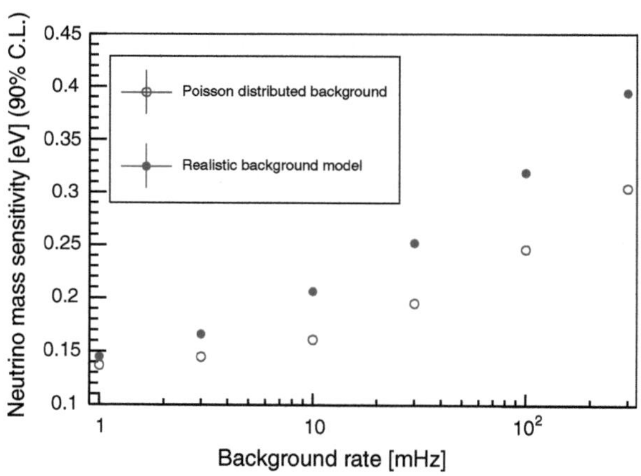

Fig. 7.24 The figure shows the sensitivity on the neutrino mass, arising from the statistical uncertainty, after 3 years of measurement time as a function of the background level. The statistical error depends on the characteristics of the background: A Poisson distributed background gives a smaller statistical uncertainty than background created by nuclear decays, which has large fluctuations correlated to individual decays. For the realistic background model a pressure of $1 \cdot 10^{-11}$ mbar and a scan time of 3 h was assumed

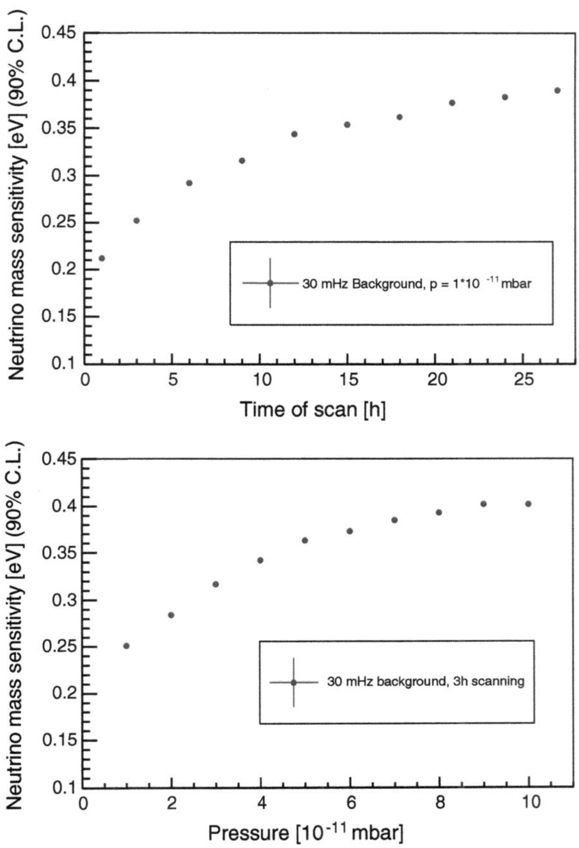

Fig. 7.25 Here the sensitivity on the neutrino mass, arising from statistical uncertainty, after 3 years of measurement time was simulated with the realistic background model for different scanning times and a fixed pressure of $1 \cdot 10^{-11}$ mbar (*top*) and different pressures and a fixed scan time of 3 h (*bottom*). As described in the text the neutrino mass sensitivity is decreased for larger scanning times, and equivalently for higher pressures

the number of secondaries of a single decay is distributed into many measurement intervals.

For the same reason the background variation depends on the actual pressure in the main spectrometer. The larger the absolute pressure in the spectrometer, the shorter the storage times of the primary electrons. Consequently, a larger pressure therefore has an analogous effect as increasing the measurement intervals.

In this regard it is also important to note that the sensitivity is improved by varying the retarding potentials during a scan in random order.

Figure 7.25 shows the statistical neutrino mass sensitivity at 90 % C.L. as a function of the scanning time and the pressure. The effect of random scanning is shown in Fig. 7.26.

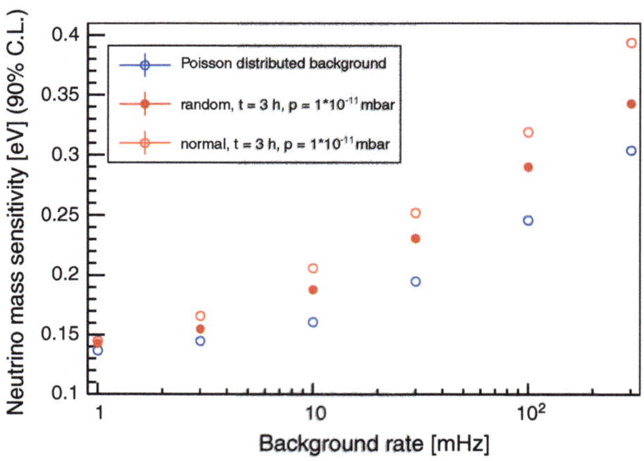

Fig. 7.26 The figure shows the neutrino mass sensitivity as a function of background rate, arising from statistical uncertainty after 3 years measurement time. Randomly choosing the potentials within one scan improves the neutrino mass sensitivity

7.5 Conclusion

Due to their intrinsic electromagnetic design, the KATRIN spectrometers act as magnetic bottles for light charged particles. A primary electron produced by a nuclear decay can thus be magnetically trapped for several hours in which it can produce up to several hundred secondary electrons.

A detailed background model based on ^{219}Rn or ^{220}Rn α-decay was developed. With this model the entire background observed at the pre-spectrometer could be explained. The excellent agreement between simulation and measurement allowed to extrapolate to the expected background at the main spectrometer.

It was shown that the nuclear decays of tritium, ^{219}Rn, or ^{220}Rn in the flux tube of the KATRIN main spectrometer can cause a background rate that largely exceeds the desired limit of 10 mHz. By increasing the amount of getter material by about 250 m while at the same time applying a liquid nitrogen cooled baffle, a background rate of about 30 mHz can be achieved.

A potentially rather dangerous characteristic of the background due to stored electrons is the associated large fluctuation of the count rate due to the correlation of a few primaries at high energy and a large secondary aftermath of low-energy secondary electrons.

A statistical analysis with a detailed background model revealed that a background of 30 mHz due to stored electrons decreases the statistical neutrino mass sensitivity to $m_\nu \approx 0.25$ eV at 90 % C.L. (compared to $m_\nu \approx 0.15$ eV 90 % C.L. assuming a Poisson distributed background of 10 mHz). This corresponds to a decrease of 60 %.

This result clearly shows the necessity of further active background reduction mechanisms. A most promising technique for background reduction which was initiated in the framework of this thesis is described in the Chap. 8.

References

1. F. Fränkle, Background Investigations of the KATRIN Pre-Spectrometer. PhD thesis, KIT, (2010)
2. F. Fränkle, L. Bornschein, G. Drexlin, F. Glück, S. Görhardt, W. Käfer, S. Mertens, N. Wandkowsky, J. Wolf, Radon induced background processes in the KATRIN pre-spectrometer. Astroparticle Phys. **35**(3), 128–134 (2011)
3. Thorium chain http://commons.wikimedia.org/wiki
4. Actinium chain http://commons.wikimedia.org/wiki
5. Uranium series http://en.wikipedia.org/wiki
6. Freedman, M.S.: Ionization by Nuclear Transitions Conference: Summer course in atomic physics, Carry-le-Rouet, France, 31 Aug 1975; Other Information: Orig. Receipt Date: 30-JUN-76, p. 18, Jan (1975)
7. S. Szucs, J.M. Delfosse, Charge spectrum of recoiling ^{216}Po in the α-Decay of ^{220}Rn. Phys. Rev. Lett. **15**, 163–165 (Jul 1965)
8. H. Tawara, Y. Itikawa, H. Nishimura, M. Yoshino, Cross sections and related data for electron collisions with hydrogen molecules and molecular ions. J. Phys. Chem. Ref. Data **19**(3), 617–636 (1990)
9. S. Görhardt, *Reduktion der durch Radon induzierten Untergrundprozesse in den KATRIN Spektrometern* (Master's thesis, KIT, 2010)
10. N. Wandkowsky, *Adiabaticity in the KATRIN spectrometers* (KATRIN Collaboration meeting, March, 2011)
11. Wolf, J.: Vacuum calculations for radon and tritium in the KATRIN pre-spectrometer and main spectrometer. Internal KATRIN report, (2011)
12. Collaboration, K.: KATRIN Design Report (FZKA Report 7090), tech. rep., KIT, (2004)
13. A. Kosmider, *DPS2-F review-preview* (KATRIN Collaboration meeting, March, 2011)
14. F. Eichelhardt, B. Bornschein, L. Bornschein, O. Kazachenko, N. Kernert, M. Sturm, First Tritium Results of the KATRIN Test Experiment Trap. Fusion Sci. Technol. **54**, 615–618 (August 2008)

Chapter 8
Active Background Reduction with Electron Cyclotron Resonance

To alleviate or potentially eliminate the background problem arising from stored electrons in the main spectrometer, a novel method needs to be implemented that actively reduces the background by removing the stored electrons. This method needs to remove both stored high- and low-energy electrons, it should not increase any other background source, and in particular it must not influence the neutrino mass measurement, i.e. the scanning process of electrons from tritium β-decay.

In the following a promising method with the potential to fulfill these requirements is presented. It is based on stochastic heating of electrons by electron cyclotron resonance (ECR). The ECR technique allows in principle to overturn the storage condition of both high- and low-energy electrons. To do so, the frequency ω of an external high frequency (HF) field is adjusted to the cyclotron frequency ω_c of the electrons in the center of the main spectrometer. For the energy interval considered here, ω_c is almost independent of the energy of the electron. Accordingly, when passing the analyzing plane, in the presence of an HF field, the electrons statistically will gain energy.

A specific advantage of the ECR technique is the short time scale for electron removal: within a short time period of 10 ms of stochastic heating, the electrons pass the resonance zone about 10^4 times. Even for moderate energy gains per passage, this leads to a significant energy increase. The basic rationale behind the ECR technique is thus to heat an electron up to an energy of about 100 keV, so that its cyclotron radius becomes larger than the radius of the main spectrometer, as visualized in Fig. 8.1.

During the short HF pulse no neutrino mass data can be taken. But since the mean ionization time of keV electrons (at 10^{-11} mbar) is of the order of 10 min and the first secondary electrons leave the spectrometer only after a few minutes after the decay, it is sufficient to apply the short HF pulse only every 10 min. Therefore the overall duty cycle of KATRIN is negligibly reduced by the HF pulsing.

In the following the interplay of a HF feed-in with the electromagnetic characteristics of the main spectrometer will be outlined. Secondly, test measurements at the pre-spectrometer proofing the feasibility of ECR as a means of background reduction

S. Mertens, *Background Processes in the Electrostatic Spectrometers of the KATRIN Experiment*, Springer Theses, DOI: 10.1007/978-3-319-01177-6_8,

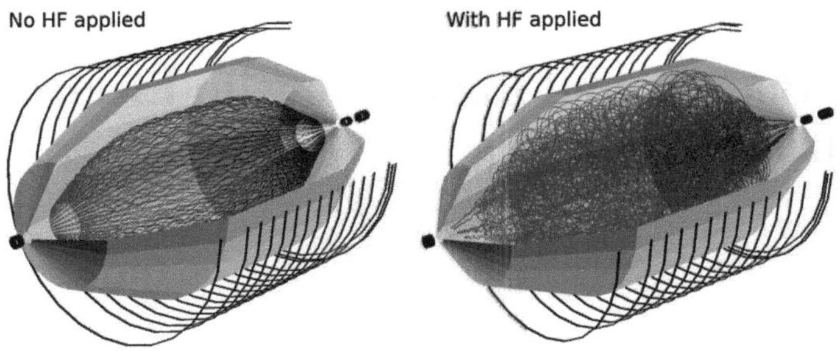

Fig. 8.1 Stored particle track (*left*). Particle track in presence of HF field (*right*). Due to the HF field the particle is hitting the electrode after less than 5 ms

will be presented. Finally, the expected background reduction efficiency at the main spectrometer will be discussed.

8.1 ECR and its Applications in Physics

The relativistic cyclotron frequency of an electron is given by

$$\omega_c = \frac{eB}{m\gamma}, \tag{8.1}$$

with e denoting the electron charge and m its mass. It describes the frequency of gyration of an electron around a magnetic field line due to the Lorentz force. By superimposing a high frequency electric field ω_{HF} tuned to the cyclotron frequency ω_c of the electron ($\omega_{HF} = \omega_c$) the electron is stochastically heated up. This phenomenon is typically used in plasma physics and condensed matter physics.

In plasma physics the ECR technique is used for plasma heating. For instance, intense, highly charged ion beams, in use at various technological fields, can be created with so called ECR ion sources. Another application is fusion technology, where the plasma heating works via so called gyrotrons. In condensed matter physics ECR can be used to determine the Fermi surface.

8.2 The Working Principle of ECR at a KATRIN Spectrometer

The cyclotron frequency of an electron (Eq. (8.1)) only weakly depends via the Lorentz factor γ on the energy. With an energy distribution of primary electrons from radon and tritium decay ranging from eV to several hundred keV the gamma factor only varies by approximately 5 %. This is exactly of the same size as the variation of the magnetic field at KATRIN along the beam axis. Therefore electrons with different energies are in resonance with the external field at some point of their trajectory. Consequently, the ECR method with fixed ω_{HF} is efficient for all energies.

The energy gain per single passage depends on the phase between the cyclotron motion of the electron and the external HF field. Applying a HF field with constant frequency ω_{HF} in a constant magnetic field will not result in any energy gain, since the phase will change periodically. At KATRIN, however, the phase between the electron cyclotron motion and the HF field changes randomly: at each transition through the resonance the energy of the electron changes. As a consequence, the electron will penetrate deeper or less deep into the magnetic mirror, which in turn leads to a phase change between the electron and the HF field.

Since the phase-space for gaining energy is much larger than the constrained phase space for loosing energy (the electron cannot have less than zero kinetic energy), a net energy gain is achieved. As it will be shown in Sect. 8.4.1, the energy gain can be further increased by applying a non-constant frequency.

8.3 Proof of Principle at the KATRIN Pre-Spectrometer

The technical implementation of applying an ECR mode during tritium scanning will be challenging in view of the rather complex system of the main spectrometer with its inner electrode system. Of course, a HF field has to comply with the integrity of all system components and will require thorough testing prior to operation in the final configuration. Therefore, the functionality of the method has first been tested at the pre-spectrometer.

8.3.1 Experimental Setup

In this section the technical implementation of the HF field feed-in to the pre-spectrometer will be presented. Furthermore, the characterization of the Krypton source, which was used as a source of high energy stored electrons will be outlined.

Technical Implementation of the HF Field

The circumference of the pre-spectrometer electrodes is 3 m. The frequency of the HF field was chosen such that its corresponding wave length is of the same order of magnitude as the electrode, allowing the formation of standing waves. The pre-spectrometer thus acts as an electromagnetic oscillating circuit, as visualized in Fig. 8.2. Standing waves lead to a resonant increase of the field strength. With a network analyzer the reflection coefficient (see Fig. 8.3) was recorded and the resonance frequencyof the pre-spectrometer could be determined. The resonance frequency

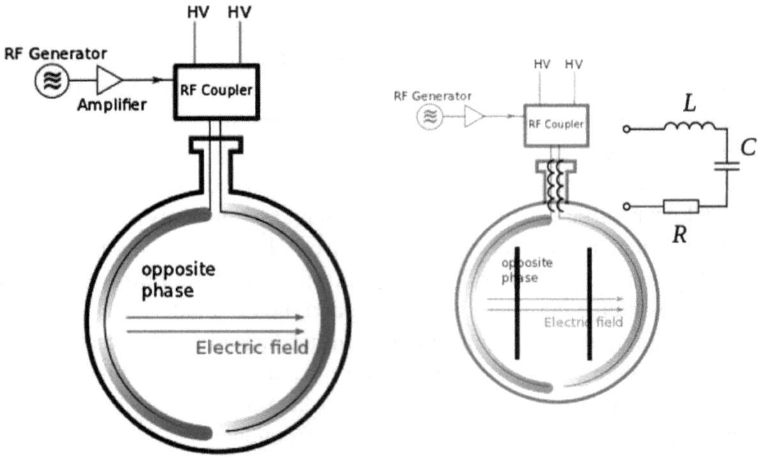

Fig. 8.2 *Left* the plot shows the cross section of the and the HF connection to the inner electrode system. The two half shells of the inner electrode system are operated opposite in phase. At 62.5 MHz a standing wave can be formed, leading to a maximal electric field strength. *Right* the figure indicates in which way the acts as an electromagnetic oscillating circuit

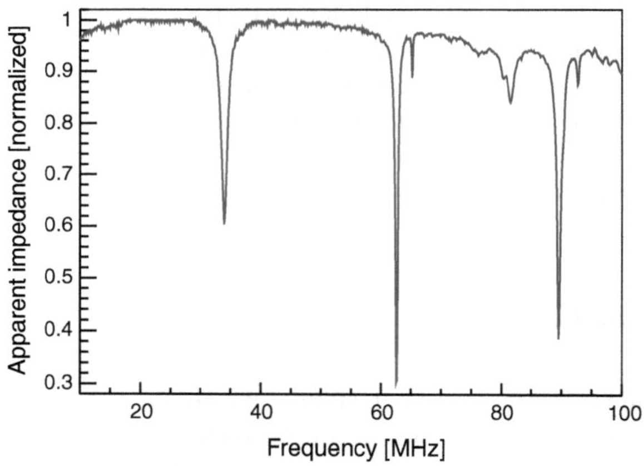

Fig. 8.3 A dip in the reflection coefficient marks a region of resonance. A sharp resonance is an evidence for a high magnification of the electric field inside the spectrometer. At a frequency of 62.5 MHz (wavelength 4.8 m) 70 % of the applied energy was absorbed by the system. This frequency was chosen as reference value

occurs at $f_{HF} = 62.5$ MHz $\left(f = \frac{\omega}{2\pi}\right)$. A cyclotron frequency of the same value, for an electron of 1 keV in the center of the spectrometer, is achieved when setting the magnetic field to about 20 G in the analyzing plane.

Figure 8.4 shows the technical implementation of the HF feed-in to the system.

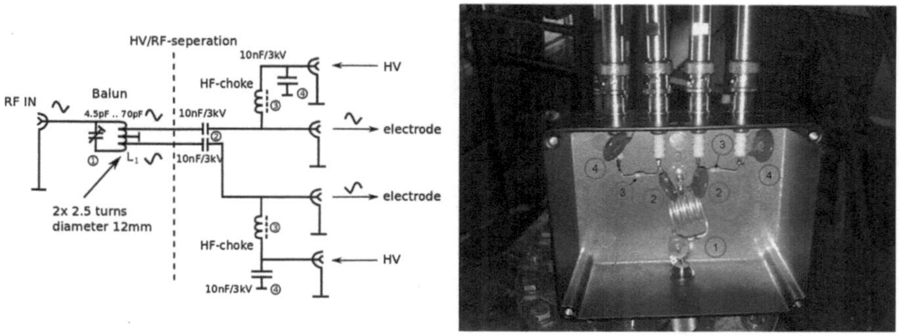

Fig. 8.4 *Left* *1* The "Balun" unit changes the single-ended signal to a differential signal (two opposite phase signals), it is constructed as a resonant circuit as shown in the schematics. *2* Capacitors to separate the HV DC component from the HF input. *3* HF choke and *4* capacitors working as LC low pass filter to prevent the HF component from entering the HV supplies. *Right* Photograph of the coupler box: superposition of HV and HF components to supply the inner electrodes

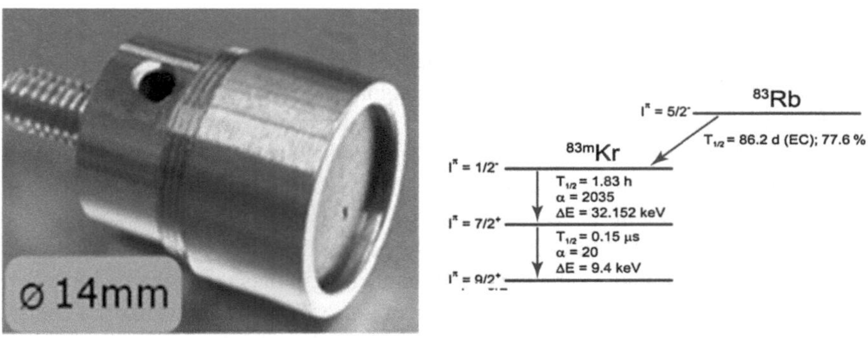

Fig. 8.5 *Left* photograph of the condensed Rubidium Krypton source. *Left* decay scheme of ^{83}Rb

Krypton as Source of Primary Electrons

To investigate the efficiency of the ECR technique a source of high energy stored electrons was required. In this work a gold-planted (Au30-1) and a platinum-planted (Pt30-2) condensed rubidium-krypton source was used (see Fig. 8.5). The condensed sources were installed at the horizontal pump port and at the e-gun position (see Fig. 3.5). Rubidium decays into an excited state of Krypton ($^{83\mathrm{m}}$Kr), which is metastable and has a lifetime of 1.83 h (see Fig. 8.5). The emanation rate of $^{83\mathrm{m}}$Kr of the Au30-1 source is $c_{\mathrm{e}}^{\mathrm{Au}}$= 11 % and $c_{\mathrm{e}}^{\mathrm{Pt}}$ = 6 % for the Pt30-2 source. Conversion and Auger electrons produced in the decay of $^{83\mathrm{m}}$Kr in the volume of the pre-spectrometer are stored due to the magnetic mirror effect.

Magnetic field dependence of background rate Since the frequency of the HF field was fixed to a value of $f_{\mathrm{HF}} = 62.5$ MHz , the magnetic field had to be adjusted so that the cyclotron frequency f_{c} of the stored electrons coincides with the external

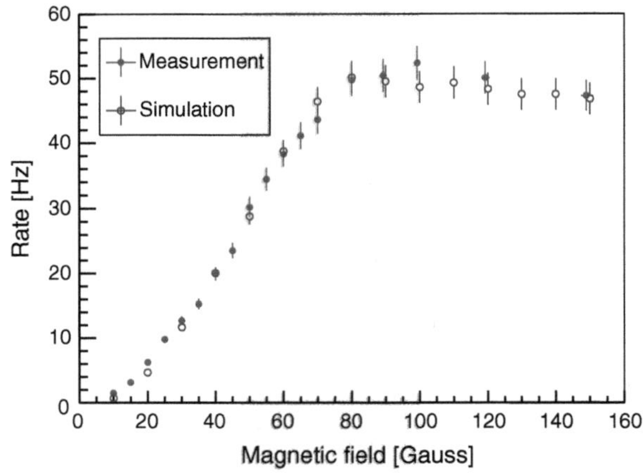

Fig. 8.6 Due to non-adiabatic effects, the rate is decreasing with decreasing magnetic fields [1]

field frequency f_{HF}, i.e. to about 20 G. To characterize the background rate created by $^{83\mathrm{m}}$Kr decays the rate was first measured at magnetic field between 10G < B < 150 G in 5 G steps (see Fig. 8.6). Interestingly, a drastic decrease of the rate for small magnetic fields was observed. This decrease of the background rate stems from non-adiabatic effects. This has been confirmed by Monte Carlo simulations, which are in very good agreement with the measurements.

Energy distribution of background In the decays of $^{83\mathrm{m}}$Kr, both conversion and Auger electrons are created. The most dominant energy lines are given in Table 8.1.

With the help of the simulated rates shown in Fig. 8.6 on can determine the energy distribution of electrons that are stored at B = 20 G. In Fig. 8.7 the energy distribution of $^{83\mathrm{m}}$Kr decay electrons is compared to the energy distribution of stored electrons at B = 20 G. This comparison shows that it is mostly the high energy electrons that are no longer stored at this magnetic field.

This statement is reinforced by the measured energy spectra at low magnetic fields. At these settings, one observes peaks at energies of (18 + 7.5) keV, (18 + 18) keV, and (18 + 30) keV which correspond to the dominant mono-energetic lines of Krypton. At low magnetic field a large number of the $^{83\mathrm{m}}$Kr decay electrons is

Table 8.1 Energy of electrons produced as consequence of ^{83m}Kr decay, [2] Only the most prominent lines are shown

Type	Probability [%]	Energy [keV]
Auger L	168.49	1.5
CE L	80.0	7.4841
CE M	13.0	9.1171
Auger K	8.63	10.8
CE K	24.8	17.8242
CE L	63.7	30.2288
CE M	10.69	31.8618

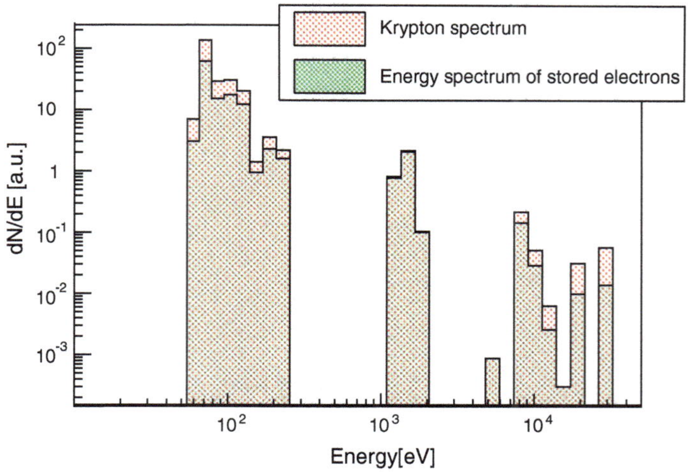

Fig. 8.7 In *red* the full kryptron spectrum is shown. At B = 20 G only a part of the created electrons is actually stored. In *green* the energy spectrum of those electrons, which are stored at B = 20 G is displayed

not stored, but immediately leaves the spectrometer. From the point of their creation to the detector they gain 18 keV net kinetic energy from the pre-spectrometer tank potential. On the other hand, at full magnetic field the energy peaks become much less prominent. Hence, at higher magnetic field the storage condition for the ^{83m}Kr decay electrons is better and the stored electrons only leave the spectrometer after they have fully cooled down. Figure 8.8 shows the measured energy spectra both at low and high magnetic field.

Krypton decay rate in the spectrometer The Krypton decay rate can be estimated from the activity of the rubidium sources $A_{\mathrm{Rb}}^{\mathrm{Au}}$ and $A_{\mathrm{Rb}}^{\mathrm{Pt}}$, the krypton emanation rates $c_{\mathrm{e}}^{\mathrm{Au}}$ and $c_{\mathrm{e}}^{\mathrm{Pt}}$, the pumping speed of the pre-spectrometer pumps λ_{TMP}, and the decay rate of $^{83m}\mathrm{Kr}$ λ_{Kr}. The krypton production and decay rate together with the pump out rate are, after some time, in an equilibrium state, i.e. the number of krypton atoms in the pre-spectrometer N_{ps} can be computed by following differential equation

$$\frac{dN_{\mathrm{ps}}}{dt} = \left(A_{\mathrm{Rb}}^{\mathrm{Au}} \cdot c_{\mathrm{e}}^{\mathrm{Au}} + A_{\mathrm{Rb}}^{\mathrm{Pt}} \cdot c_{\mathrm{e}}^{\mathrm{Pt}}\right) - (\lambda_{\mathrm{Kr}} + \lambda_{\mathrm{TMP}})\, N_{\mathrm{ps}} = 0. \tag{8.2}$$

The activity of the Au30-1 source and the Pt30-2 source during the measurements were $A_{\mathrm{Rb}}^{\mathrm{Au}} = (27.6 \pm 3)$ kBq and $A_{\mathrm{Rb}}^{\mathrm{Pt}} = (40.6 \pm 4)$ kBq, respectively. The corresponding emanation rates are $c_{\mathrm{e}}^{\mathrm{Au}} = (11 \pm 2)$ % and $c_{\mathrm{e}}^{\mathrm{Pt}} = (6 \pm 2)$ %. The relative pumping rate of the two pumps at the pre-spectrometer (Ebara and Leybold pump) are estimated to be $\lambda_{\mathrm{TMP}} = (0.074 \pm 0.015)\ \mathrm{s}^{-1}$. Finally, the decay rate of ^{83m}Kr is known to be $\lambda_{\mathrm{Kr}} = \frac{\ln(2)}{1.83}\ \mathrm{h}^{-1}$.

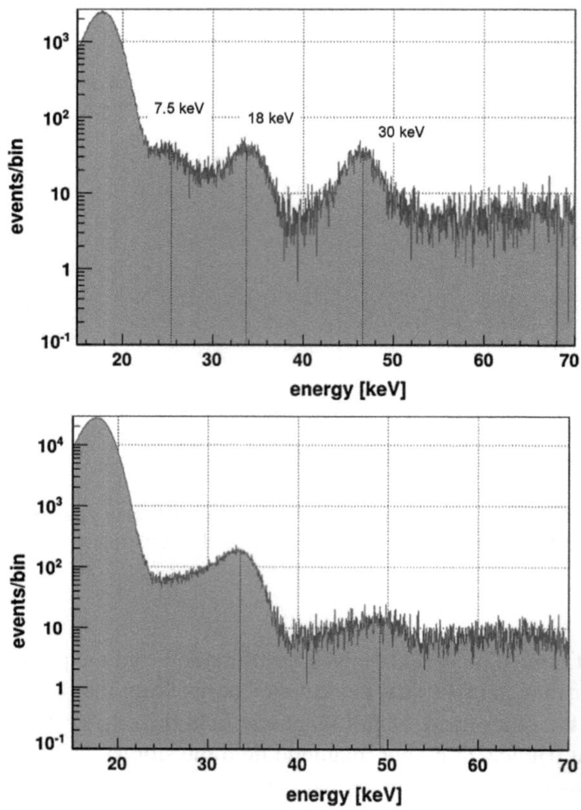

Fig. 8.8 *Top* The measurement parameters are magnetic field in analyzing plane: B = 20 G, tank potential $U_{\text{tank}} = -18$ kV and inner electrode potential $U_{\text{inner}} = -18.5$ kV. At low magnetic field the electrons of the Krypton lines at 7.5, 17.8 and 30 keV are no longer stored. On their way to the detector they gain 18 keV due to the retarding potential. The peaks in the energy spectrum are not exactly at the line position plus 18 keV. This might be due to the fact that the electrons are stored for some time, in which they loose some of their energy and only then are released from the trap due to non-adiabaticity. *Bottom* The magnetic field in analyzing plane is increased to B = 159 G. At high magnetic field the storage conditions are much better. The peak at roughly 30 + 18 keV has almost disappeared, i.e. event the 30 keV electron are stored until they are cooled down. There is still a peak at roughly 18 + 18 keV which is likely due to pile-up effects of cooled down electrons in the energy region of interest (ROI) of $15 < E < 21$ keV

After rearranging Eq. (8.2) and inserting the values given above one finds:

$$A_{\text{Kr}} = \lambda_{\text{Kr}} N_{\text{ps}} = \frac{\left(A^{\text{Au}}_{\text{Rb}} \cdot c^{\text{Au}}_{\text{e}} + A^{\text{Pt}}_{\text{Rb}} \cdot c^{\text{Pt}}_{\text{e}}\right) \cdot \lambda_{\text{Kr}}}{\lambda_{\text{TMP}} + \lambda_{\text{Kr}}} = (7.7 \pm 2.1)\text{s}^{-1}. \quad (8.3)$$

To extrapolate to the decay rate in the sensitive volume of the spectrometer one has to take into account that the detector only surveys 25 % of the flux tube and the flux tube volume covers only 28.5 % of the total pre-spectrometer volume. Hence, the observable decay rate is

$$A_{\mathrm{Kr}}^{\mathrm{obs}} = A_{\mathrm{Kr}} \cdot 0.285 \cdot 0.25 = (0.55 \pm 0.15)\mathrm{s}^{-1} \quad (8.4)$$

A different approach to estimate the decay rate is via comparison of the simulation with measurement shown in Fig. 8.6. The number of electrons reaching the detector per $^{83\mathrm{m}}$Kr decay was simulated for different magnetic fields allowing to determine the relative rate at each magnetic field. Since the number of $^{83\mathrm{m}}$Kr decays is independent of the magnetic field, the number of $^{83\mathrm{m}}$Kr decays per second $A_{\mathrm{Kr}}^{\mathrm{sim}}$ is a single free parameter in the simulation that can be fitted to match the data.

With a χ^2 fit the best fit value is found at $A_{\mathrm{Kr}}^{\mathrm{sim}} = 0.13 \pm 0.0019\,\mathrm{s}^{-1}$. The result is of the same order of magnitude as the result obtained with the vacuum considerations. In the latter it was assumed that the complete emanated rate enters the pre-spectrometer. In this assumption a fraction that might stick to the vessel surface on the way from the source position to the volume is neglected which might explain the difference of the results.

8.3.2 Measurement Results

Coming back to the measurements we recapitulate that the goal there was to show that a background reduction can be achieved and in particular that the reduction is due to the effect of ECR. Secondly, the measurements should prove that the reduction can be achieved by short HF pulses. Finally, it was to be demonstrated that no other background source is enhanced due to the HF feed-in.

Resonance Effect

Since the HF field has a fixed frequency of $f_{\mathrm{HF}} = 62.5$ MHz, the cyclotron frequency of the electrons needs to be tuned to the HF field, by adjusting the magnetic field correspondingly as visualized in Fig. 8.9. To proof that the background reduction is actually due to the ECR effect, the magnetic field was varied below and above the value B = 21.5 G where the resonance is expected. A resonant reduction is expected when the external HF field coincides with the cyclotron frequency of the electrons in the center of the spectrometer.

Figure 8.10 shows the reduction factor of the background rate as a function of the magnetic field. A clear resonant reduction is observed at 21.5 G, as expected for ECR. The width of the curve can be explained by the inhomogeneity of the magnetic field in the pre-spectrometer. If the magnetic field is too low for resonance to occur in the center of the spectrometer, there are still resonant regions further away from the center. However, if the magnetic field is above the resonant value in the center,

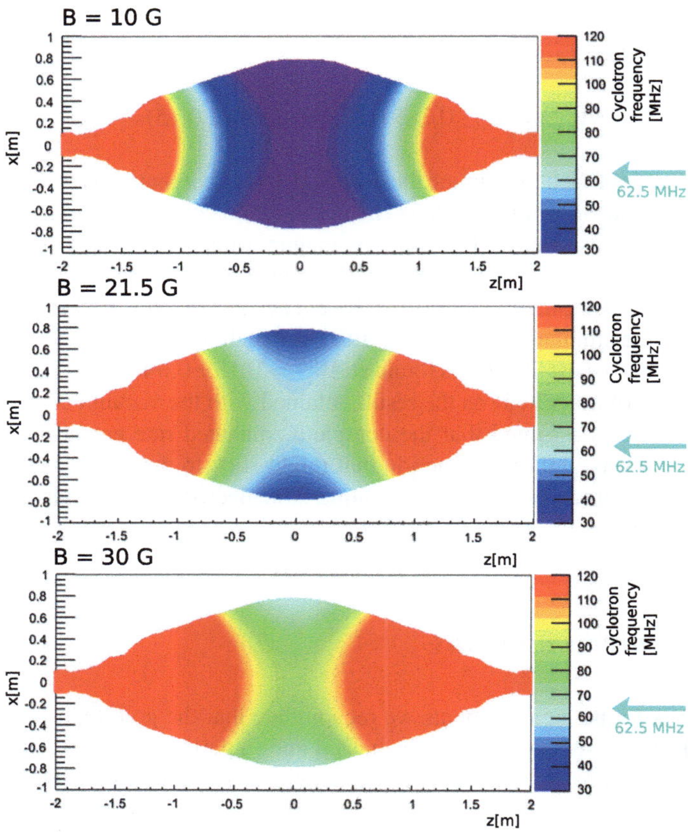

Fig. 8.9 The *color coding* indicates the cyclotron frequency for a 1 keV electron. In the *light blue* area the cyclotron frequency coincides with the frequency of the external HF field. At magnetic field strength lower than B = 21.5 G this occurs only at the entrance and exit of the spectrometer. As the electron passes through these regions very quickly, no large reduction is expected. At B = 21.5 G, the cyclotron frequency of the electrons in the center is just equal to the HF field frequency, and consequently a large reduction is expected. For magnetic fields larger than B = 21.5 G no resonance is met

it is above the resonance in the entire spectrometer. This explains the asymmetric shape of the curve.

Effect of HF Pulsing

It is evident that during the HF pulse no neutrino mass measurement can be performed. Therefore the HF pulsing must be as short as possible. Corresponding simulation revealed that the removal time of stored electrons is considerably shorter than one second. To experimentally test the effect of short HF pulses the rate increase

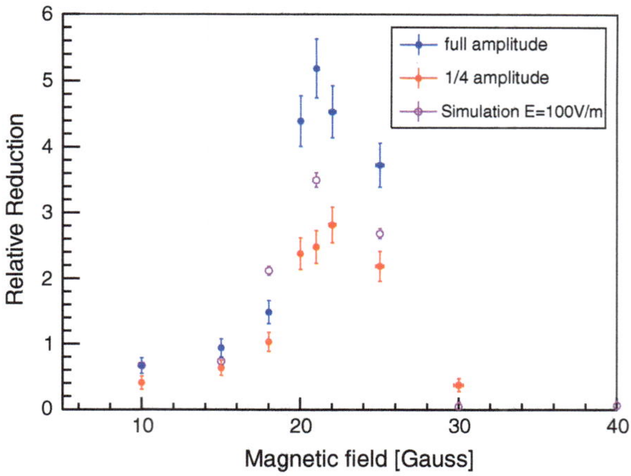

Fig. 8.10 The relative reduction is defined as $\frac{\text{Rate}_{\text{off}} - \text{Rate}_{\text{on}}}{\text{Rate}_{\text{on}}}$. As expected the maximal reduction is observed at B = 21.5 G. The measurement time at each point was about 5 h. The measurement was performed for two HF field amplitudes. The HF field amplitude, was controlled by adjusting the power feed into the system. The actual field inside the spectrometer, can be determined by comparison to the simulation. The setting at full field amplitude and magnetic fields of B > 25 G lead to a detector breakdown, therefore no data could be taken. The data convincingly show the expected resonance pattern of the ECR technique

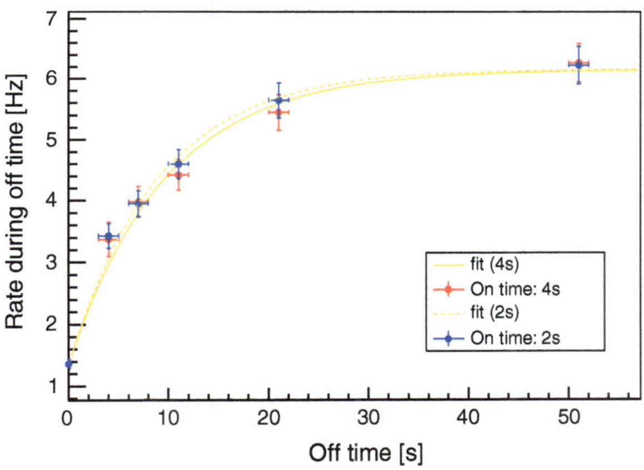

Fig. 8.11 To test the effect of pulsed ECR the rate was measured at a magnetic field of B = 21.5 G, for a condition, where the HF field was switched on for 2 s (4 s) and then switched off for $t_{\text{off}} =$ 5–50 s. For each t_{off} setting a measurement of about 10 h was performed. In the plot the average rate during t_{off} is shown

after the pulse was measured for two different pulse lengths. Indeed an inverse exponential increase following a $(1 - e^{-\frac{t}{\tau}})$ scaling was found, as expected for a background arising from stored electrons. Figure 8.11 shows that the reduction of rate is independent of whether the pulse duration is 2 or 4 s. For purely technical

Table 8.2 Background rate measured without krypton source

HF setting	Background rate [mHz]
No HF	11.5 ± 0.9
With HF	6.2 ± 1.17

To test the influence of the HF field on other background sources a long-term measurement without the krypton source at B = 21.5 G was performed, both with continuous HF field and without HF field. This measurement was analyzed with the pulse shape analysis (PSA) and time correlation filter (TCF) [3]

reasons no shorter pulses could be tested. However, we can thus conclude that the removal time is smaller than 2 s.

Influence on Other Background Sources

A well known background mechanism is related to cosmic muons hitting the spectrometer vessel thereby releasing electrons from the inner surface (see Chap. 5). These electrons are electrically and magnetically shielded from the sensitive volume of KATRIN. To test whether a HF field has any influence on this shielding a measurement without krypton source was performed. Table 8.2 shows that the intrinsic background rate was not enhanced due to the HF field. We can thus conclude that the shielding works even in the presence of a HF field.

8.4 Expected Background Reduction at the Main Spectrometer

To validate and study in detail the concept of a stochastic heating by the ECR process at the main spectrometer, Monte Carlo simulations with KASSIOPEIA were performed.

The HF field was implemented in a first order approximation as a sinusoidal function of the form

$$\vec{E}\,(x,t) = \vec{E}_0\,(x)\sin\left(\omega\,(t)\cdot t\right) \tag{8.5}$$

where $\vec{E}_0\,(x)$ is the amplitude of the HF field, ω is the frequency and t is the time.

The amplitude $\vec{E}_0\,(x)$ is chosen to be oriented perpendicularly to the beam axis, as only this component of the field is able to increase the transversal energy of the electrons. The frequency was assumed to be either stationary or sweeped through in steps. Typical frequencies for a reference setting of the magnetic field in the analyzing plane are 8–10 MHz.

In an initial step, the functionality and validity of the implemented routines in Kassiopeia were tested by comparing the numerical result to an analytical calculation [4, 5]. In particular, the energy gain of a low-energy electrons in a constant magnetic field under the influence of a high frequency electric field was considered. The agreement between the numerical result produced by KASSIOPEIA and the analytic solution is of the order of 10^{-10}, which is of sufficient precision.

The goal of the simulations was to optimize the frequency settings, as well as to test the efficiency of ECR with low electric field amplitudes, and finally to investigate the effect of ECR on electrons of different energies.

8.4.1 *Optimizing the Frequency Settings*

An important fundamental aspect of the simulations presented below was to find out the optimal frequency setting of the HF field. As mentioned, due to the non-constant magnetic field in the main spectrometer, electrons of all energies will be in resonance with the HF field at some point of their trajectory. However, it will be shown that by changing the frequency within a predefined frequency band will enhance the resonance effect and considerably improve the efficiency of stored particle removal.

First Tests of ECR in a Constant Magnetic Field

First, the influence of the ECR technique on relativistic electrons with a starting energy of $E^{start} = 5$ keV in a constant magnetic field of $B = 3$ G was investigated. The corresponding cyclotron frequency is $f = 6$ MHz, which corresponds to a cyclotron period of $T = 1.67 \cdot 10^{-7}$ s.

The first interesting observation is that running the HF field with a constant frequency will change the energy of the electron periodically as shown in Figs. 8.12 and 8.13. Sweeping the frequency turns out to be a more efficient way to heat electrons. The frequency sweep can be implemented based on a step function or randomly. Moreover, the duration of a step and the number of steps were investigated. The simulations showed that the energy change of the electron by ECR is most efficient, if

- the time period of constant frequency of the HF field is of the order of a few times the cyclotron period.
- Within the frequency band many different frequencies occur. It is thus advantageous if either the frequency is changed randomly, or changed in many small steps.

ECR in the KATRIN Main Spectrometer

In a non-constant magnetic field configuration as in the main spectrometer the cyclotron frequency of the electrons changes along their trajectory. Therefore a HF field with fixed frequency might be sufficient to remove electrons of all energies.

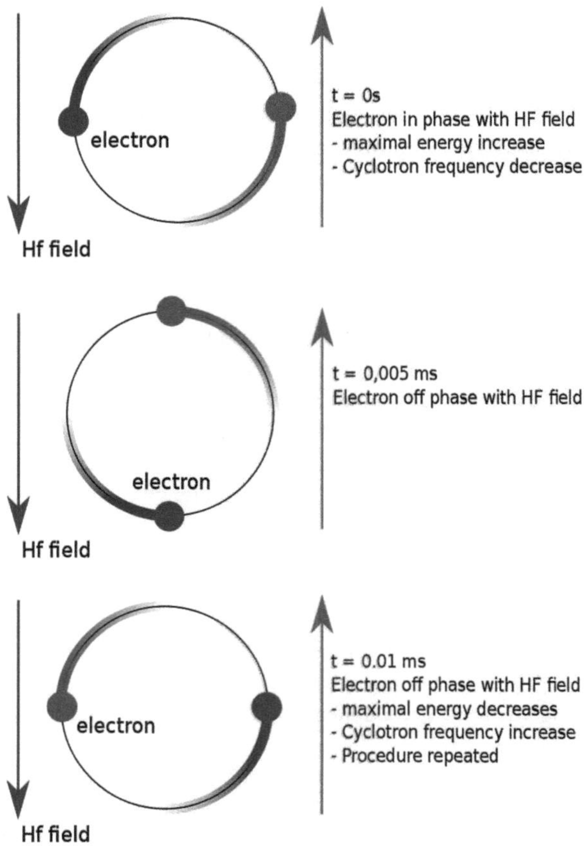

Fig. 8.12 If the electron is in resonance with the HF field it gains energy, thereby increasing its Lorentz γ-factor, which in turn reduces its cyclotron frequency. After some time the electron is off phase with the HF field meaning its energy is maximally reduced and its cyclotron frequency is increased. Consequently, in a constant HF field an oscillatory energy change is observed

However, by varying the frequency one can still improve the background reduction efficiency. This was investigated by considering electrons from tritium β-decay under the influence of the ECR technique, first with a constant frequency of the HF field, then with sweeping frequency.

To interpret the results of the simulations in a reasonable way, one needs to compare the simulation of stored electrons both with and without HF field. Even without HF field a large number of electrons created by tritium β-decay in the volume of the spectrometer, immediately hit the walls of the main spectrometer due to their large cyclotron radius, another non-negligible fraction of decay electrons leaves the spectrometer axially due to non-adiabatic effects. The distribution of exit conditions of stored electrons without HF field is shown in Fig. 8.14.

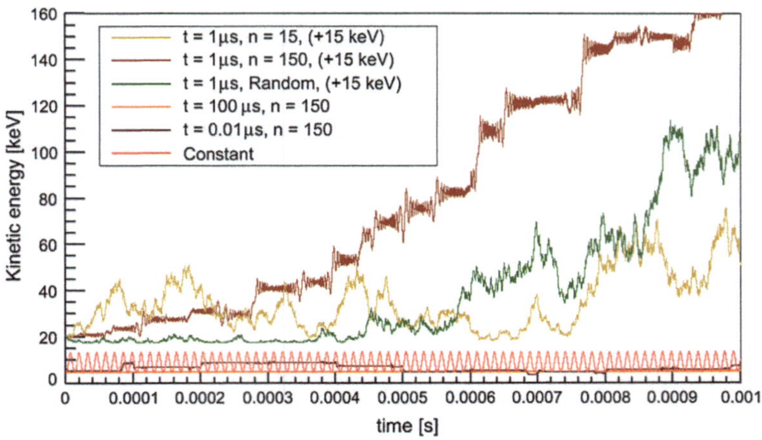

Fig. 8.13 The figure shows the kinetic energy of an electron in a constant magnetic field of $B = 3$ G in presence of a HF field in sweeping or constant mode as a function of time. In the sweeping mode the frequency was varied between $f_{min} = 7$ MHz and $f_{max} = 8.5$ MHz, the constant frequency was set to $f_{const} = 8.28$ MHz. The HF amplitude was chosen to be $E_0 = 100$ V/m. In the sweeping mode the frequency is sweeped through in n steps with an individual time step t. The frequencies are either varied stepwise or randomly (in the latter case the number of steps is not fixed). The calculation shows that the HF sweeping time steps t needs to be of the order of a full cyclotron turn. If the time steps are either too short or too long (*orange* and *dark brown line*) no energy increase will be observed. A constant frequency leads to a periodic energy change (see also Fig. 8.12). The three upper lines have been shifted by 15 keV for better visibility

Table 8.3 ECR simulation settings for constant frequency

ECR paramter	Setting
Mode	Constant
Frequeny	$f = 9.750$ MHz
HF amplitude	$E_0 = 100$ V/m

The constant HF frequency ω_{HF} is chosen to correspond to the frequency of a low-energy electron (1 eV) in the lowest magnetic field. With a fixed frequency ω_{HF}, defined in this way, electrons of all energies eventually hit the resonance frequency. In the sweeping frequency mode, the frequency is changed between 8 and 10 MHz (corresponding to the cyclotron frequency of a 1 eV electron and a 100 keV electron in the analyzing plane). In doing so, the frequency is changed in 200 steps with the duration of each step being 1 µs. Both for constant and sweeping HF field the amplitude is chosen to be $E_0 = 100$ V/m.

The comparison of the two runs shows that in the case of constant frequency a fraction of 8 % of the electrons can not be removed within 0.5 s. In case of a frequency band all electrons could be removed (see Fig. 8.14). However, even if not all electrons can be removed within one HF period this is not a generic problem, as the subsequent HF pulses are expected to remove a significant fraction of these electrons. The measuring interval in between subsequent HF pulses is most likely

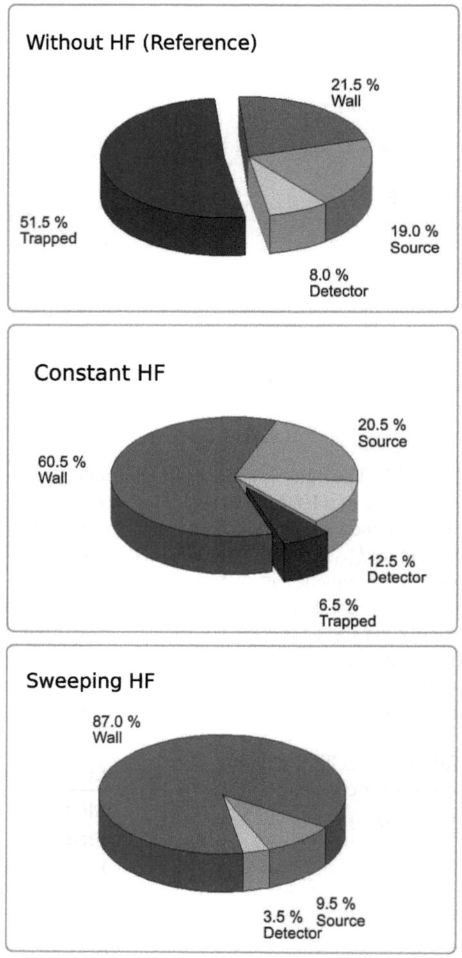

Fig. 8.14 The labeling of the pie charts refers to the exit condition of the primary electrons from tritium β-decay for the three scenarios being considered. **Detector** Electron leave the spectrometer axially in the direction of the detector. **Source** Electron leaves spectrometer axially in source direction. **Wall** Electron hits the wall. **Trapped** Electron is stored for more than 0.5 s. The ECR settings for constant and sweeping frequency are given in Tables 8.3 and 8.4, respectively. The figures demonstrate the high efficiency of the ECR technique, in particular in the sweeping mode with a reduction of stored electrons by 100 %. The number of primary electrons reaching directly the detector is a negligible background contribution

unaffected by these electrons which are "parked" at high energies, due to the energy dependent decrease of the cross section for ionizing collisions.

8.4.2 Optimizing the Amplitude of the HF Field

For obvious reasons with respect to the integrity of the wire electrode the lowest possible amplitude of the HF field is preferable. In the following the effect of a HF field with sweeping frequency and amplitudes from $E_0 = 10-200$ V/m on electrons from tritium β-decay in the main spectrometer is investigated. For all E_0 all electrons

Table 8.4 ECR simulation settings for sweeping frequency

ECR Paramter	Setting
Mode	Sweeping
Frequeny	$f_{\rm min} = 8$ MHz
	$f_{\rm max} = 10$ MHz
Time of step	$\Delta t_{\rm step} = 1\ \mu$s
Number of steps	$n_{\rm step} = 200$
HF amplitude	$E_0 = 100$ V/m

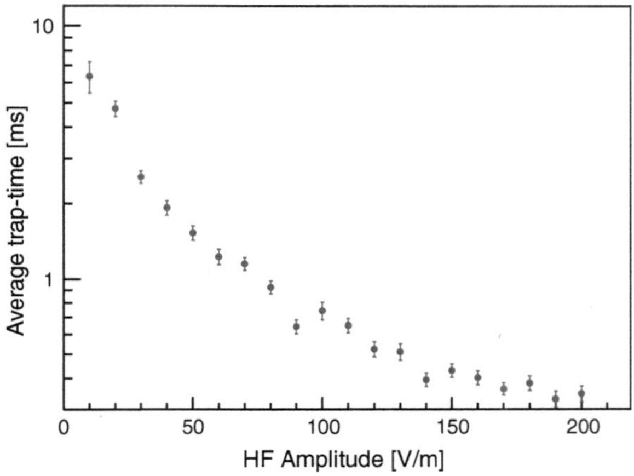

Fig. 8.15 The higher the amplitude the faster the stored electrons can be removed. At $E_0 = 10$ V/m the average removal time is still exceedingly fast with less than 10 ms. For this figure 10,000 tritium β-decay electrons were simulated under the influence of ECR in the main spectromter. The ECR simulation settings are given in Table 8.4

could be removed. Even for an amplitude as low as $E_0 = 10$ V/m the removal time was only ~10 ms as shown in Fig. 8.15.

8.4.3 Effect of an HF Field on Electrons of Different Energies

As electrons of different energies are stored in the main spectrometer, the effect of a HF field as a function of the initial electron energy was investigated. Again, the following background sources are expected:

- Electrons following radon α-decay of up to 500 keV.
- Electrons following tritium β-decay of up to 18.6 keV.

The scientific goal here is of course to demonstrate that the ECR method is capable to remove all energies from eV up to keV. Indeed, the simulations showed that both low and high energy electrons can be removed completely. Figures 8.16 and 8.17

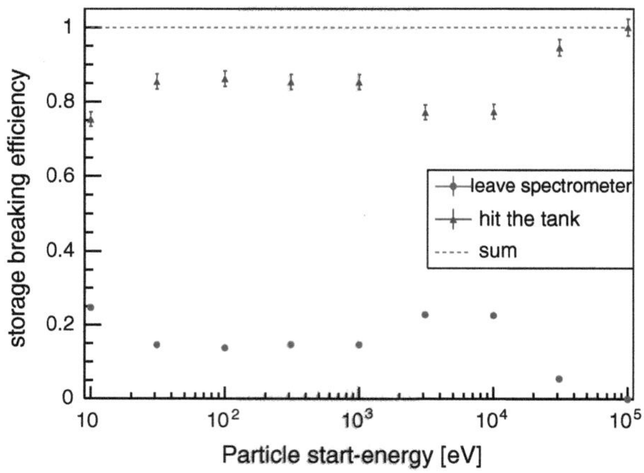

Fig. 8.16 The figure shows the exit conditions of electrons starting in the main-spectrometer as a function of their start-energy. The overall removal efficiency 100 %. Most electrons (~80 %) will hit the vessel walls due to an increase of the transversal energy by ECR, while a minority (~20 %) leaves the spectrometer as a result of non-adiabatic motion

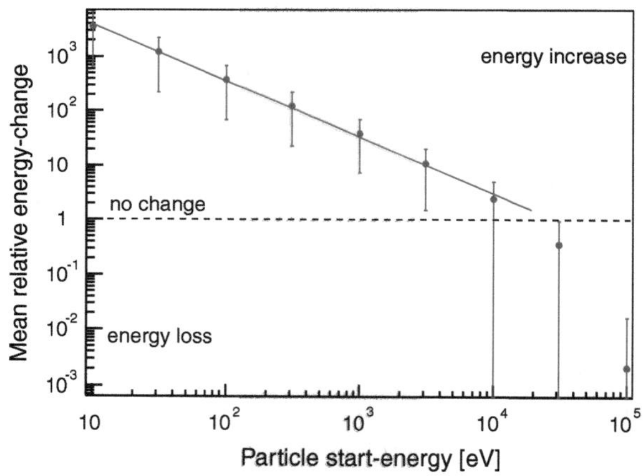

Fig. 8.17 This figure shows the ratio of final electron energy (defined by the exit condition) to the initial particle energy. While low-energy electrons in the tens of eV range have to be boosted by factors $> 10^3$, the required boost-factor decreases for increasing particle energy, as the exit condition is typically met at fixed transversal energy. This characteristic behavior generates the observed simple power-law spectrum

give information of the corresponding exit conditions and the relative energy change as a function of the energy of the stored electrons. For these figure 22,000 electrons of 10 eV < E_{fix} < 100 keV were simulated under the influence of ECR in the main spectrometer. The ECR simulation settings are given in Table 8.4.

8.5 Conclusion

In this chapter it was shown that stochastic heating with the ECR technique is a most promising tool to reduce the background arising from stored electrons following nuclear decays.

Extensive measurements at the pre-spectrometer have given experimental proof of the basic functionality of the ECR method. A resonant reduction of the background rate was observed when the cyclotron frequency of the stored electrons was tuned to the frequency of the external HF field. It could be confirmed that short HF pulses of $\Delta t < 2$ s are sufficient to achieve the maximal background reduction factor of 7. Finally, it could be shown that the HF field does not affect the magnetic and electric shielding against the muon induced background.

The corresponding simulation demonstrate the extraordinary potential of the ECR technique to remove electrons over a wide energy range. Since the removal time is of the order of ms and the mean ionization time of a keV electron is of the order of 10 min in the main spectrometer, the duty cycle of KATRIN is not significantly affected by making use of HF pulsing. It could be shown that a HF field with sweeping frequency and even a low amplitude of only 10 V/m (assuring the integrity of the wire electrode) is sufficient to reduce the background entirely.

References

1. N. Wandkowsky, Adiabaticity in the KATRIN spectrometers. in *KATRIN Collaboration meeting*, March 2011.
2. S.-C. WU, Nuclear data sheets for A = 83. Nucl. Data Sheets 92(4), 893–1028 (2001)
3. F. Fränkle, Background Investigations of the KATRIN pre-spectrometer. PhD thesis, KIT (2010)
4. E.H. Holt, R.E. Haskell, Foundations of plasma dynamics. Macmillan (1965)
5. R. Geller, Electron cyclotron resonance ion sources and ECR plasmas. Institute of Physics Pub. (1996)

Chapter 9
Conclusion

In the framework of this thesis a full Monte Carlo simulation software for KATRIN, the KASSIOPEIA package was successfully developed in collaboration with the KATRIN simulation task group. The software comprises several particle event generators, such as a sophisticated radon and krypton decay event generator, which was implemented for the purpose of this work. A major module of the software, developed in the frame of this work, is the precise and fast tracking of particles in electromagnetic fields. By comparing simulation results to several background observations KASSIOPEIA could be validated.

Based on the successful implementation of KASSIOPEIA the influence of non-axially symmetric magnetic fields on the magnetic shielding against muon-induced background was studied first. The investigations have revealed that magnetic materials present in the KATRIN experimental hall and deformations of the air coil system will cause rather large disturbances of the magnetic shield. In particular, they can lead to radial drift velocities of up to 300 m/s for secondary electrons from the inner surface of the vessel. A Monte Carlo simulation with KASSIOPEIA has revealed that $3 \cdot 10^{-4}$ of the electrons experience a sizable radial drift allowing them to reach the sensitive spectrometer volume and thus create background.

In a next step of this work, ultra-precise electromagnetic field calculations with KASSIOPEIA revealed the existence of a very small Penning trap at the end ring of the ground electrode in the pre-spectrometer with a cross sectional area of only $0.75\ \text{cm}^2$ and a depth of $U_{\text{trap}} = 0.9$ kV. The installation of a modified electrode which eliminated the trap confirmed that this trap was actually responsible for a kHz background. In this work a phenomenological model was developed to explain how such a tiny Penning trap can cause a background level exceeding the design limits by 5 orders of magnitude. The model is based on the creation of UV-photons in the trap and has been confirmed by comparisons of theoretical results with corresponding measurements.

The major focus of this work was targeted at the modeling of a novel background source observed at the pre-spectrometer. Again a detailed physics model based on the hypothesis of ^{219}Rn and ^{220}Rn α-decay in the volume of the spectrometer was

S. Mertens, *Background Processes in the Electrostatic Spectrometers of the KATRIN Experiment*, Springer Theses, DOI: 10.1007/978-3-319-01177-6_9,

developed. Accordingly, a detailed radon event generator was implemented into KASSIOPEIA and extensive simulations were able to fully reproduce the measured results. The background model allowed to describe central background features such as event topology, time duration of electron storage as well as other parameters. This validated model was thus used to extrapolate the background issue to the situation at the main spectrometer, where in addition to radon α-decay also tritium β-decay will occur in the sensitive volume.

The corresponding simulations have revealed that based on the original design values the reference background level of 10 mHz is exceeded by about two orders of magnitude. As a consequence, a passive shielding against ^{219}Rn realized by a nitrogen cooled baffle will be installed between the spectrometer and the getter pump, which is the strongest ^{219}Rn emanator. Furthermore, to reduce the partial pressure of tritium in the main spectrometer the pre-spectrometer now features the option to be used as an additional tritium pump. This can be achieved by installing additional getter material around its inner electrode system. With both these improvements a background level of 30 mHz can be reached. However, due to non-poissonian fluctuations stemming from multi-keV stored electrons following nuclear decays it was found that the neutrino mass sensitivity of KATRIN is reduced by up to 50 %.

As a solution to the problem arising from stored electrons in the main spectrometer a novel method based on stochastic heating of the stored electrons by Electron Cyclotron Resonance (ECR) was developed. Corresponding measurements at the pre-spectrometer, using krypton as a source for stored electrons, could prove that by applying a High Frequency (HF) field exactly tuned to the cyclotron frequency of the stored electrons will lead to a resonant reduction of the observed background rate. This is due to the significant increase in transversal energy. Furthermore, the measurements have shown that short HF pulses of <2 s are sufficient to maximally reduce the background. Finally, long-term measurements proved that no other background source is increased by the HF field.

In addition it could be shown by simulations that in the case of the main spectrometer all stored electrons over a wide energy band from 1 eV < E < 100 keV can be removed within less than 10 ms even for low HF amplitudes of $E_0 = 10$ V/m. The short removal time implies negligible measurement time losses to KATRIN, due to the tiny duty cycle of the ECR periods, while low HF amplitudes are necessary to assure the integrity of the wire electrode system. Both simulation and measurement results makes the ECR technique a very promising tool for background reduction.

In conclusion it can be summarized that this work has allowed to eliminate various sources of background at the pre-spectrometer from an initial level of several kHz down to a sub-mHz level. This huge reduction of background by more that 6 orders of magnitude is an essential cornerstone of achieving the design neutrino mass sensitivity. In doing so it has been essential to develop detailed physics models for muon-induced background, Penning traps, and the generation of electrons following nuclear decays. However, only by developing and validating the extensive KASSIOPEIA package it was possible to successfully compare experimental results with Monte Carlo data. KASSIOPEIA now forms a unique tool for accompanying upcoming measurements at the main spectrometer. At present, all sources of background in the spectrometers down to the mHz regime have been identified in this work. This will allow the KATRIN experiment to fully explore its physics potential in the hunt for the absolute mass scale of neutrinos.

Appendix A
Optimization of the Spectrometer Transmission Properties with the Air Coil System

The air coil system surrounding the main spectrometer vessel consists of two subsystems: The Earth Magnetic field Compensation System (EMCS) and the Low Field Coil System (LFCS). The EMCS is targeted to compensate the earth magnetic field both in vertical and horizontal direction. The LFCS on the other hand is used for fine tuning of the magnetic field close to the analyzing plane.

The LFCS consists of 15 concentric rings with individually adjustable currents, allowing to define the magnetic field in the analyzing plane B^{ana} (the design value of $B^{ana} = 3 \cdot 10^{-4}$ T). Moreover, the LFCS allows for rather flexible tuning of the magnetic field over a large area of the main spectrometer. This fact can be used to improve the transmission properties for the β-decay electrons.

The EMCS consists of 16 loops for vertical earth magnetic field correction ($B^{v} = 45 \cdot 10^{-6}$T), and 10 loops for horizontal earth field correction ($B^{h} = 3 \cdot 10^{-6}$ T) (see Fig. A.1).

A.1 Electromagnetic Design Criteria

With the help of the air coil system the electromagnetic properties of the KATRIN spectrometer section can be optimized with respect to following criteria.

First, the radial dimensions of the magnetic field lines of the flux tube must be smaller than the radial dimensions of the tank and its inner electrode system. If the flux tube area is too large the electron count rate would be reduced since electrons are guided to electrode surfaces. Moreover, the background will be increased as electrons emitted from the electrode surface would be guided directly to the detector. For a given magnetic field strength in the WGTS source there is a minimal magnetic field in the analyzing plane assuring that the flux tube fits into the tank. Figure A.2 shows the impact of the air coil system on the flux tube.

Second, the transmission condition must be satisfied. When electrons propagate from the entrance of the main spectrometer to its center the decrease of the magnetic

S. Mertens, *Background Processes in the Electrostatic Spectrometers of the KATRIN Experiment*, Springer Theses, DOI: 10.1007/978-3-319-01177-6,

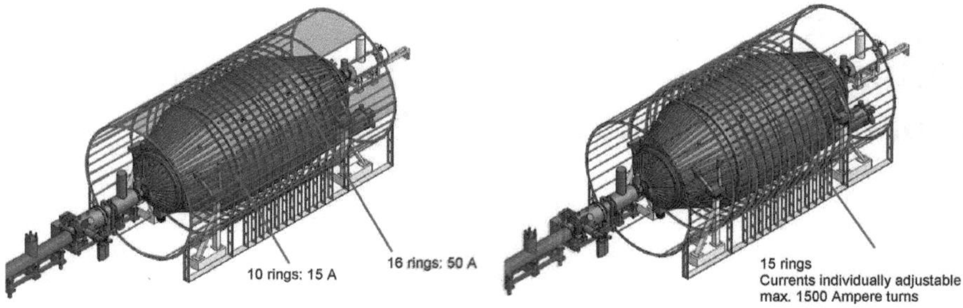

Fig. A.1 Layout of the air coil system components. *Left* Earth magnetic field compensation system (EMCS). *Right* Low field coil system (LFCS)

field must be precisely adjusted to the increase of the electric retarding potential. Any deviation from the optimal setup could lead to a too early retardation of signal electrons and to the storage of secondary electrons. In Sect. A.2 it will be explained how the settings of the LFCS can optimize the transmission condition.

Third, the magnetic field in the analyzing plane should be as homogeneous as possible. Once the transmission condition is satisfied, the magnetic and electric field in the analyzing plane define the measured spectrum. Since both electric potential and magnetic field are not fully homogeneous in the analyzing plane, a multi-pixel detector is used at KATRIN to compensate this. With this technique several sub-areas of the analyzing plane can be treated separately in the final analysis. To each of the sub-areas defined by radial and azimuthal position a particular magnetic field strength and electric potential is assigned. It is evident of course that a more homogeneous magnetic field results in less spread of the magnetic field within one sub-area.

As it has been pointed out above, the magnetic field should be as axially symmetric as possible. This requirement is linked to the well-known fact that the magnetic field serves as a shield against electrons emitted from the surface of the electrodes. This shielding can only be guaranteed with a perfectly axially symmetric field (see Chap. 5).

A task of the electromagnetic design is to identify the best strength of the magnetic field in the analyzing plane. On the one hand lowering the magnetic field in the analyzing plane increases the energy resolution, i.e. the transmission function gets sharper. On the other hand, the higher the magnetic field, the better it serves as magnetic shielding. The latter effect arises due to the fact that then the flux tube is further away from the tank. Electrons emitted from the electrode surface need to travel larger distances in order to reach the flux tube. Moreover, since the distance of the field lines and the inner electrode surfaces is larger the danger of violating the transmission condition is reduced and the occurrence of small Penning traps within the flux tube is less likely. Finally, in a stronger magnetic field the influence of azimuthal components of the magnetic field is smaller, therefore the conservation of axial symmetry is better.

Another important constraint comes from the requirement that the motion of electrons is adiabatic. Therefore the magnetic field along the magnetic field lines should not decrease too fast. Owing to the due to the exceedingly large dimensions of the main spectrometer tank, this condition is rather easily fulfilled for signal electrons.

A.2 Two Magnetic Field Minima Configuration

Based on the criteria defined above the LFCS has been designed in a way that adjustments of the coils allow for two local magnetic field minima of about $2 \cdot 10^{-4}$ T. These minima are created at about $z = \pm 3.5$ m from the analyzing plane as shown in Fig. A.4. This configuration improves both the transmission condition for electrons and the homogeneity of the magnetic field in the analyzing plane. The magnetic field in this layout is more homogeneous along the beam axis and consequently also more homogeneous in radial direction (see Fig. A.5).

The transmission condition is defined in the following way: The starting energy of an electron which lead to a longitudinal energy of zero in the analyzing plane, is called transmission energy. The transmission condition is fulfilled if:

- an electron with a starting energy lower than the transmission energy is reflected,
- an electron with a starting energy higher than the transmission energy is transmitted,
- and along all field lines the longitudinal energy of an electron starting with the transmission energy is zero in the analyzing plane and positive everywhere else.

In order to fulfill the transmission condition the magnetic field has to be adjusted to the electric potential in the correct way: As the increasing electric potential decreases the longitudinal energy of the electrons, the decreasing magnetic field has to convert transversal energy into longitudinal energy, accordingly.

The transmission condition can be mathematically derived. In a first step an expression for the transmission energy E_{trans}, as defined above, needs to be found. For this purpose the kinetic energy E_{kin} and the transversal energy $E_{\perp}$ at the position $\vec{p}$ along a magnetic field line are expressed as a function of the total starting kinetic energy E_{start} the starting angle Θ_{start}, the electric potential $U(\vec{p})$ and the magnetic field $B(\vec{p})$

$$E_{\text{kin}}(\vec{p}) = E_{\text{start}} + q(U_{\text{start}} - U(\vec{p})) \quad \text{(A.1)}$$

$$E_{\perp}(\vec{p}) = E_{\text{start}} \cdot \sin^2 \Theta_{\text{start}} \cdot \frac{B(\vec{p})}{B_{\text{start}}}. \quad \text{(A.2)}$$

The transmission condition then demands for an electron starting with the transmission energy

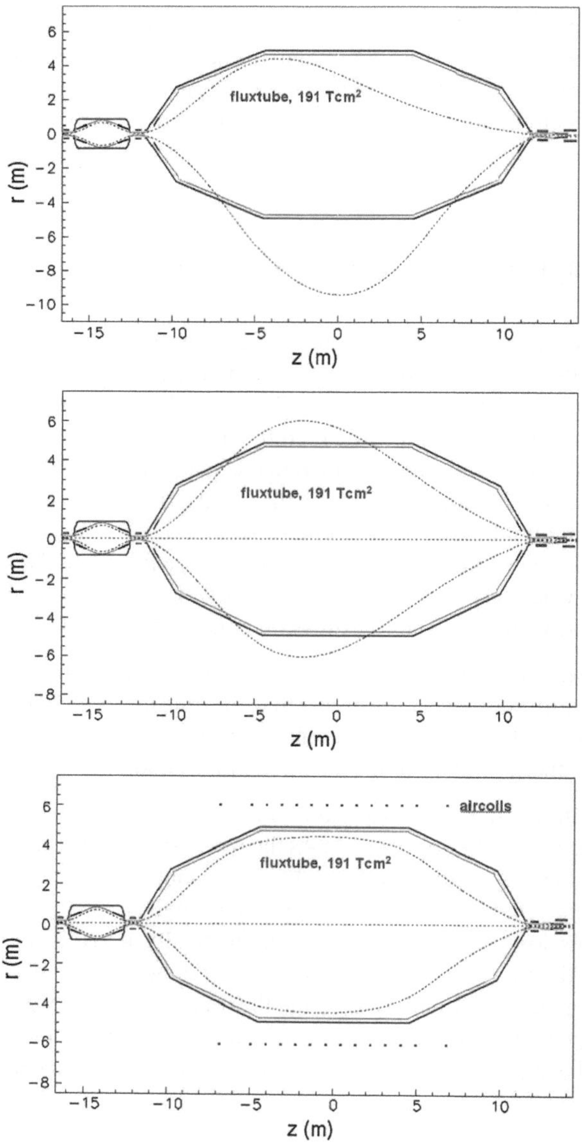

Fig. A.2 Impact of air coil system on flux tube geometry. *Top* Flux tube without air coil system. *Middle* Flux tube geometry with EMCS. *Bottoms* Flux tube geometry with optimized EMCS and LFCS settings

$$E_{\parallel}(z=0) = 0 \tag{A.3}$$

$$\rightarrow E_{\text{kin}}(z=0) = E_{\perp}(z=0), \tag{A.4}$$

where z is the axial position of the electron. Using this relation between kinetic energy and transversal energy we find the transmission energy to be

$$E_{\text{trans}} = \frac{-q(U_{\text{start}} - U(z=0))}{1 - \sin^2 \Theta_{\text{start}} \cdot \frac{B(z=0)}{B_{\text{start}}}}. \tag{A.5}$$

Finally the transmission condition can be derived by considering the longitudinal energy $E_{\parallel}$ of an electron starting with the transmission energy. The $E_{\parallel}(\vec{p})$ must be larger than zero for each $z \neq 0$.

$$\begin{aligned} E_{\parallel}(\vec{p}) &= E_{\text{kin}}(\vec{p}) - E_{\perp}(\vec{p}) \\ &= E_{\text{start}} \left(1 - \sin^2 \Theta_{\text{start}} \cdot \frac{B(\vec{p})}{B_{\text{start}}}\right) + q(U_{\text{start}} - U(\vec{p})) \\ 0 &< E_{\text{trans}} \left(1 - \sin^2 \Theta_{\text{start}} \cdot \frac{B(\vec{p})}{B_{\text{start}}}\right) + q(U_{\text{start}} - U(\vec{p})) \\ \rightarrow q \cdot \frac{U(\vec{p}) - U_{\text{start}}}{U(z=0) - U_{\text{start}}} &< \frac{B_{\text{start}} - \sin^2 \Theta_{\text{start}} \cdot B(\vec{p})}{B_{\text{start}} - \sin^2 \Theta_{\text{start}} \cdot B(z=0)} \end{aligned} \tag{A.6}$$

If the electric potential rises too fast (from entrance to center) as compared to the decrease of the magnetic field an electron might be reflected unintendedly. Moreover, if an electron is produced in the volume of the spectrometer it is accelerated towards the ends by the electric potential. However, at the same time its longitudinal energy is transformed into transversal energy. If the decrease of the electric potential is too slow (from the center to the entrance/exit) compared to the rise of the magnetic field, this particle can be trapped.

Furthermore, satisfying the transmission condition assures that the transmission function is only depends on the magnetic field in the region of the source and the analyzing plane. If the longitudinal energy of an electron starting with the transmission energy is zero not only in the analyzing plane but elsewhere, there might be several reflection points. Hence, the knowledge of the magnetic field and electric potential in a large region around the analyzing plane, namely in all points of reflection, is needed.

In the following three cases of the interplay between magnetic field and electric potential are considered. Figure A.3 visualizes the three scenarios.

- **Case I: Non-homogeneous electric potential:** This case is only mentioned for pedagogic reasons as it is not feasible at KATRIN. Here we consider an electric potential configuration in which the potential has its negative maximum (-18 kV) in the analyzing plane and is strongly decreasing (to less negative values) to both sides along the beam axis, as shown in Fig. A.3. The reflection point of an electron starting with the transmission energy is well defined by the shape of the electric potential without strong constraints on the magnetic field.
- **Case II: Homogeneous electric potential with a single magnetic field minimum:** Since the main spectrometer vessel is on high voltage, the electric potential is rather homogeneous along the beam axis and rapidly increasing at the

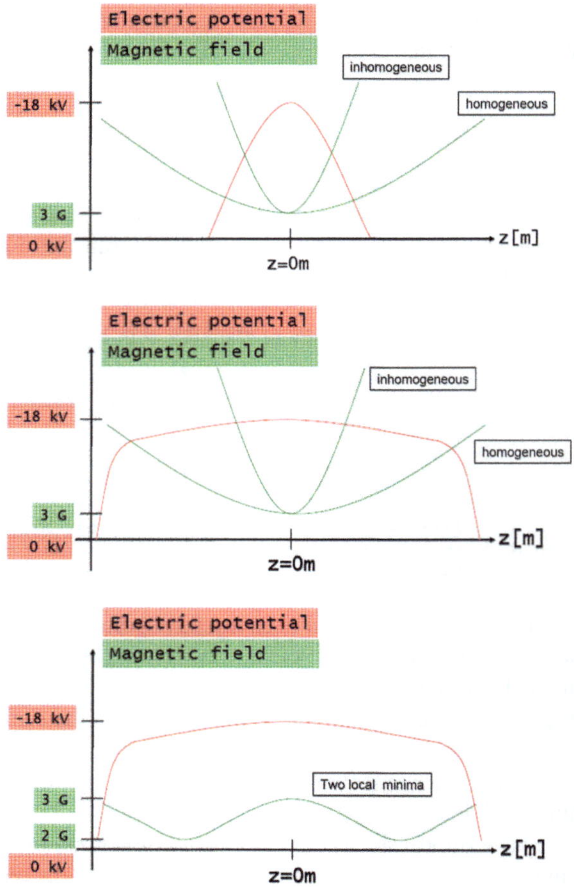

Fig. A.3 Transmission properties in three different electromagnetic field configurations. *Top* Case I considers a rather inhomogeneous electric retarding potential. In this configuration which is not feasible at KATRIN the electric potential alone assures to fulfill the transmission condition. Case II and III assume a realistic electric potential which is rather homogeneous along the beam axis since the entire vessel is set to high voltage. *Middle* If the magnetic field has only a single minium (Case II) the field has to be very homogeneous in order not to violate the transmission condition. *Bottom* With two local magnetic field minima (Case III) the transmission condition is easily fulfilled even in the case of a very homogeneous electric potential

entrance (exit). This configuration of the electric potential is preferable since it improves the electric potential homogeneity in radial direction. However, in this configuration the transmission condition is not easily satisfied. To assure that the longitudinal energy of an electron starting with the transmission energy is zero in the analyzing plane and positive everywhere else, the magnetic field must decrease very fast as the electric potential increases fast at the entrance (exit). However, once the electric potential is only slowly increasing the magnetic field must not decrease further to assure the minimal longitudinal energy is in the analyzing plane.

- **Case III: Homogeneous electric potential with two local magnetic field minima:** To avoid the rather critical situation described in Case II, a magnetic field configuration with two local minima can be chosen. In this case the magnetic field is minimal before the electric potential reaches its negative maximum, which prevents the longitudinal energy of the electron to drop to zero in this region. Close to the analyzing plane the magnetic field increases which causes a decrease

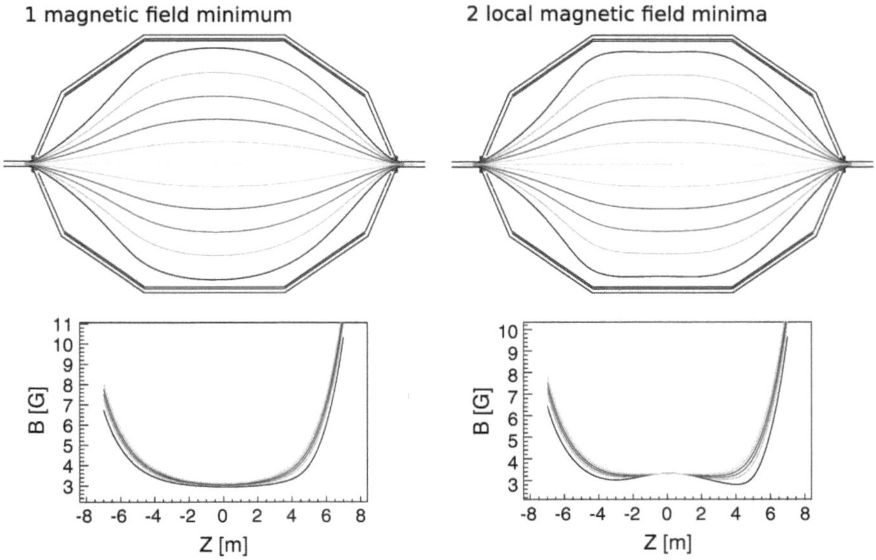

Fig. A.4 Comparison of the two different magnetic field configurations. *Left* Single magnetic field minium. *Right* Two local magnetic field minima

of the longitudinal energy of an electron flying towards the analyzing plane. Therefore this special shape of the magnetic field assures that the longitudinal energy of an electron starting with the transmission energy is minimal in the analyzing plane and nowhere else. Since the fixed configuration of the electric potential of KATRIN does not automatically assure to fulfill the transmission condition, this configuration might be preferable.

A drawback of the magnetic field configuration with two local minima is that particles starting in the region of the magnetic field minimum under a certain minimal angle with respect to the magnetic field line are stored [1]. This angle Θ_{min} depends on the depth of the magnetic field minimum defined by B_{min} and B_{max}.

$$\Theta_{min} = \arcsin\sqrt{\frac{B_{min}}{B_{max}}}. \tag{A.7}$$

The electric potential is approximately constant within the region of a local magnetic field minimum therefore only the magnetic field is responsible for the trap. In this case the trapping condition is independent of the starting energy. A magnetic field in the analyzing plane of $B_{max} = 3\,G$ and $B_{min} = 2\,G$ in the region of the local minimum results in a minimal angle of $\Theta_{min} = 56°$. However, the ionization energy for H_2 is maximal at about 100 eV and electrons with starting energies of 100 eV already have a large trapping probability due to the magnetic mirror effect present in the main spectrometer (see Sect. 7 and Fig. 7.15).

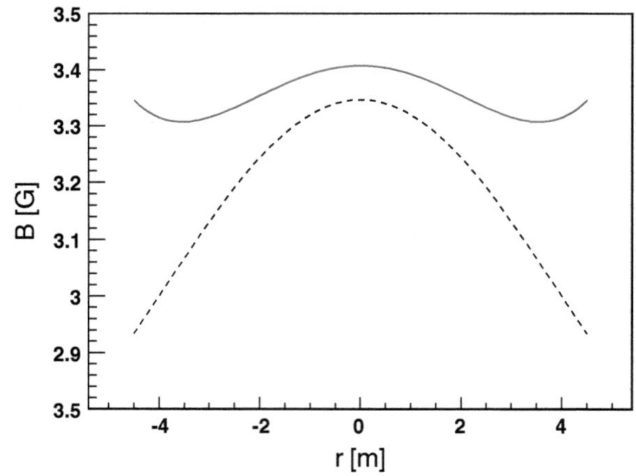

Fig. A.5 Radial homogeneity of the magnetic field in the analyzing plane for the two different magnetic field configurations. *Red solid line* Two local minima, *dashed black line* single global minimum

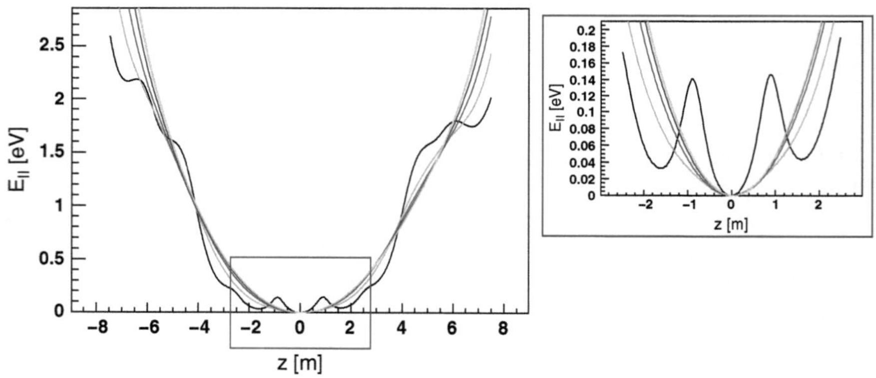

Fig. A.6 Longitudinal energy of electrons starting with transmission energy. Along different field lines shown in Fig. A.4 for two local magnetic field minima. The longitudinal energy only drops to zero only in the analyzing plane. The most critical situation occurs along the outermost field lines

To conclude, the LFCS and EMCS co-designed in this work will allow to optimize the transmission characteristics of the spectrometer, while at the same time allow to study background phenomena in detail.

Appendix B
Comparison of Transmission Measurements at the Pre-spectrometer with Kassiopeia Simulations

The MAC-E-Filter principle (see Sect. 3.2) in use both at the main and pre-spectrometer, requires an adiabatic motion of the β-decay electrons along their full trajectory from the source to the detector. Non-adiabaticity occurs most likely for high-energy electrons in low magnetic fields (see Sect. 7.3.1).

As described in Sect. 3.4.3 the setup with pre- and main spectrometer at full potential forms a major Penning trap between the two spectrometers. Stored electrons in the trap produce an unacceptably high background rate. To mitigate this problem arising from the Penning trap two methods are foreseen: (a) a wire scanner could be installed in the center of the Penning trap periodically removing the stored electrons, and (b) the pre-spectrometer could be set to zero or very low voltage to avoid the creation of the trap. With the second solution the kinetic energy of the electrons passing through the pre-spectrometer is no longer retarded by the potential, i.e. they pass a rather low magnetic field region of $B_{\mathrm{min}} = 156\,\mathrm{G}$ with full kinetic energy. This situation might lead to non-adiabatic effects.

To test the degree of adiabaticity of electrons with high surplus energies in the pre-spectrometer corresponding test measurement were performed [2, 3]. In the course of these measurements the transmission rate of electrons produced by an electron gun (e-gun) was measured for different retarding potentials (i.e. surplus energies of the electrons). Non-adiabatic effects would disturb the magnetic collimation of the electrons, and consequently, electrons with enough kinetic energy to be transmitted would be reflected, which would lead to a reduction of the count rate. Even though adiabaticity calculations early on assigned the absence of non-adiabatic effects even for large surplus energies, the measurement did show a disturbing reduction of the count rate for large surplus energies.

To explain this phenomenon in detail, a full simulation of the measurement with Kassiopeia was performed. A very good agreement between measurement and simulation was found. The key to understand the measurement result was the integration of electron interactions in the silicon detector (performed by the simulation module Kess [3], described in Sect. 4.3.4 and the implementation of a detailed model of the e-gun in the simulation. It could be shown that the reduction of the count rate can be explained mainly by detector backscattering rather than non-adiabatic effects in the

S. Mertens, *Background Processes in the Electrostatic Spectrometers of the KATRIN Experiment*, Springer Theses, DOI: 10.1007/978-3-319-01177-6,

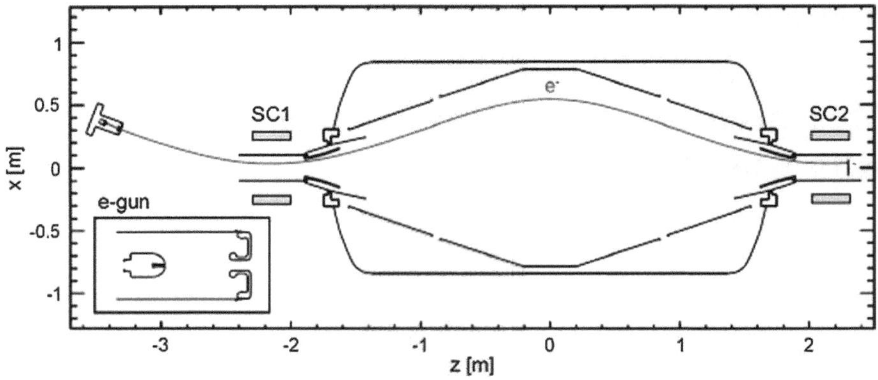

Fig. B.1 Electromagnetic simulation setup [3]. The figure shows a cross section of the pre-spectrometer geometry used for the simulations. The inlet shows a close up view of the e-gun. The position of the e-gun in this picture is off-axis. SC1 and SC2 present the superconducting solenoids. Inside of SC2 the silicon pixel detector is placed

spectrometer. The agreement between simulation and measurement is an important validation of Kassiopeia.

In the following the measurement and simulation settings and comparisons as well as a detailed explanation of the reduction of the transmission of electrons with high surplus energies will be given.

B.1 Measurement Settings

A view of the experimental setup is given in Fig. B.1. A measurement of the transmission was carried out with the following magnetic field and e-gun settings:

- The magnetic field in the center of the magnet was set to
  - $B = 4.5$ T,
  - $B = 2.3$ T.
- For both magnet setting the e-gun angle to beam axis was set to
  - $\alpha = 0°$
  - $\alpha = 15°$
  - $\alpha = 19°$

For each setting the retarding potential was varied between -17.5 kV $< \mathrm{U_{ret}} <$ -0.5 kV in 1 kV steps. For each potential the rate of transmitted electrons from the e-gun was measured. The energy region of interest (ROI) was chosen to be 15 keV $<$ E $<$ 21 keV. All measurement, but the B $= 4.5$ T with $\alpha = 0°$, show a reduction of the rate for high surplus energies, as shown in Fig. B.4.

B.2 Phenomenological Explanation of Measurement Result

Electrons hitting the Si-surface of the detector with an energy of 18 keV and an incident angle of 0° have a backscattering probability of 20 %. This probability is increased for higher angles. The backscattered electrons can be reflected again by the magnetic field (if the detector is behind the maximal magnetic field), or by the retarding potential, or finally by the e-gun potential. Backscattered electrons with high surplus energies have a large probability to be backscattered only by the e-gun potential, since it is likely that they did not loose enough energy in the detector to be reflected by the retarding potential.

The electrons return to the detector on a time scale of ns whereas the DAQ shaping time is of the order of a few μs. Consequently, they are still detected as a single event. In the following it will be shown that not all backscattered electrons will eventually end up at the detector again.

Measurement at $B = 4.5$ T and $\alpha = 0°$
At full magnetic field $B = 4.5$ T and zero angle $\alpha = 0°$ of the e-gun no rate reduction was observed, as expected for full adiabatic motion. For high surplus energies the electrons loose not enough kinetic energy in the detector to be reflected by the retarding potential, instead they are reflected by the e-gun potential. All reflected electrons reach the detector again and therefore no rate reduction is observed.

Measurements with off-axis e-gun ($\alpha = 15°$ **and** $\alpha = 19°$)
In the case of an off-axis e-gun the magnetron motion of the electrons needs to be taken into account. The $\vec{E} \times \vec{B}$-drift and the $\vec{\nabla}|\vec{B}| \times \vec{B}$-drift lead to an azimuthal drift of the electrons in addition to their motion along the beam axis (see Fig. 5.4). Consequently, as the electron is backscattered from the detector it does not take exactly the same path back to the e-gun, but its trajectory is azimuthally rotated due to the magnetron drift. If the e-gun is off-axis the backscattered electron can instead of being again reflected again by the e-gun potential, hit the e-gun ground blind. Hence, this electron is lost and does not reach the detector again. This process is visualized in Fig. B.2. Combining the phenomenon of backscattering and taking into account the magnetron motion of the electron explains the observed reduced count rate.

Measurement at $B = 2.3$ T and $\alpha = 0°$
As above described, the reduced rate can be explained for off-axis e-gun positions, however, at $B = 2.3$ T a reduced rate was measured even for the e-gun being on-axis. To understand this result yet another phenomenon needs to be taken into account. In a combination of a low magnetic fields and high electric field the motion of electrons is likely to be non-adiabatic. Inside the e-gun set-up the electrons are accelerated to 18 keV on a very short path of about 30 mm. Consequently, a high electric field is present inside of the e-gun. At a low magnetic field of $B = 2.3$ T (corresponds to $B \approx 1$ T at the e-gun) the motion of the backscattered electrons in the area of the high electric field is no longer adiabatic. This non-adiabaticity leads to a change of the particle's angle with respect to the field line, i.e. the polar angle of the incoming electron is not the same after its reflection. When the electrons return from the e-gun

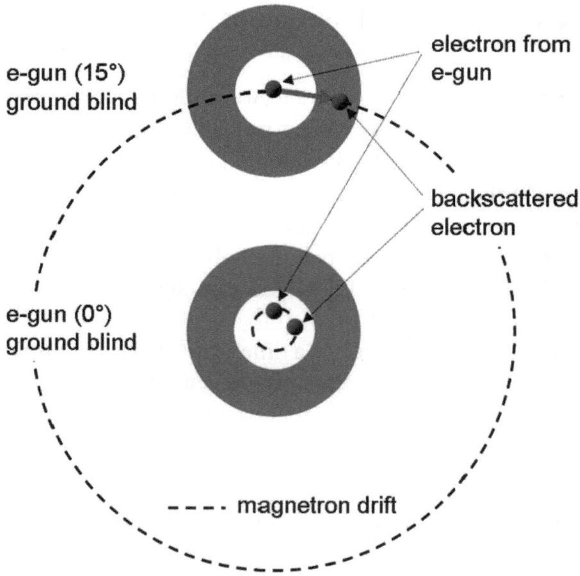

Fig. B.2 View of the e-gun ground blind from the direction of the detector. The backscattered electron performs a magnetron motion, indicated by the *dashed line*. In case the e-gun in on-axis, the backscattered electron is reflected by the e-gun potential and eventually reaches the detector again. In case the e-gun is off-axis the magnetron motion of the electron guides the electron onto the ground blind of the e-gun

potential they have a high probability to be magnetically reflected by the solenoids and thus be trapped between magnet and e-gun potential. This mechanism is visualized in Fig. B.3. Only a very small angle range would allow the trapped electrons to escape from the trap. Therefore, the trapped electrons can be considered as lost for the measurement, which explains the observed reduced rate at high surplus energies.

B.3 Comparison with Monte Carlo Simulation

To obtain a more quantitative understanding of the observations, a detailed Monte Carlo simulation with Kassiopeia was performed. The exact calculation mode of KTrack (see Sect. 4.3.2) allows for taking into account non-adiabatic effects, while the module Kess (see Sect. 4.3.4) handles the simulation of backscattering at the detector. Finally, very precise electric field calculation methods (see Sect. 4.3.3) based on Legendre polynomial expansion and elliptic integrals allow for a detailed model of the e-gun and the pre-spectrometer potentials.

Simulation Settings
The generic electromagnetic simulation setup is shown in Fig. B.1.

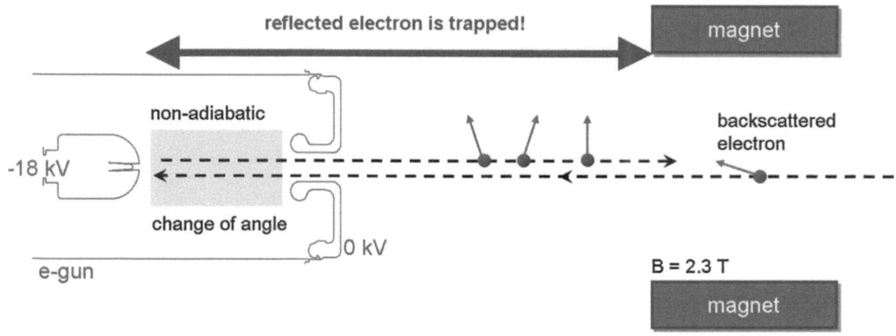

Fig. B.3 Side view of the e-gun and the solenoids at the e-gun side of the prespectrometer. Some of the backscattered electrons fly back into the e-gun where they are reflected again. The electric field inside the e-gun is very high, since the distance from the e-gun tip to the ground blind is only about 3 mm. In case the magnetic field is rather low, the motion of the electrons inside of the e-gun is non-adiabatic. Consequently, their angle to the magnetic field line is changed and they have a large chance to be magnetically reflected by the solenoid and be trapped between magnet and e-gun potential subsequently

The electrons are uniformly started from a disk of a diameter of 1 mm, representing the e-gun gold tip. The angular distribution of the e-gun is based on measurements in [4] and is given by $\Theta = \arcsin(R)$ and $\Phi = 2\pi R$, with the random number $R \in [0, 1]$. The electrons are created with a uniform energy distribution between $0\,\text{eV} < E < 2\,\text{eV}$.

The detector simulation Kess was used to account for the energy loss of the electrons in silicon, to include the detector response as well as the angular and energy distribution of the backscattered electrons. In particular, the dead layer thickness was set to $\lambda = 150$ nm and the energy resolution was fixed to $\Delta E_{FWHM} = 3.5$ keV.

An electron backscattered at the detector is typically reflected and reaches the detector again. Since, as mentioned, the DAQ shaping time is slower than the time-of-flight of the electrons, the total deposited energy E_{dep} of an electron in the sensitive detector volume is computed by adding up the deposited energies at each detector hit. The detected energy thus is simulated by folding E_{dep} with a Gaussian of the width ΔE_{FWHM}. Finally, the same energy ROI as in the measurement is chosen in the simulation.

To account for electron losses, the following exit conditions were chosen: The electron hits the e-gun ground blind, the electron is trapped (more than 20 turns), or the electron has less than 100 eV kinetic energy.

Comparison and Conclusion

To compare the measurement with the simulation the rates were normalized so that the sum of all counts for all settings of the retarding potential U_r amounts to unity (or 100 %). Figure B.4 shows the simulated and measured count rates. A very good agreement for all runs is found.

This result shows that the reduction of the count rate is not due to potential non-adiabatic transmission properties of the pre-spectrometer. The simulations

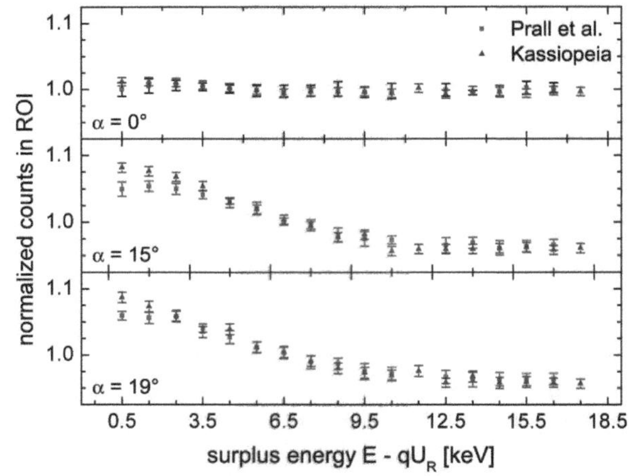

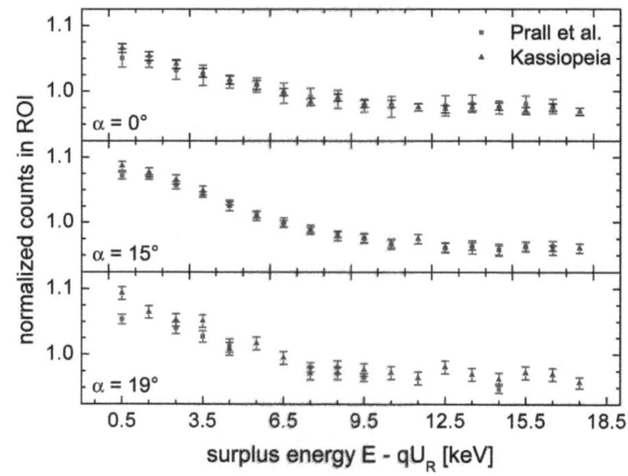

Fig. B.4 Comparison of the Kassiopeia results (this work and [64]) with the experimental data [?].*Top* B = 4.5 T. *Bottom* B = 2.3 T

reveals that detector backscattering in combination with magnetron motion and non-adiabaticity inside of the e-gun are responsible for the observed rate reduction. In the final KATRIN setup it is therefore possible to run the pre-spectrometer at zero potential in order to avoid the creation of a Penning trap between the two spectrometers. This is of major importance for the long-term neutrino mass measurements with minimum background levels.

Appendix C
Alternative Methods of Removing Stored Electrons

Stored electrons in the keV-energy range following nuclear decays of radon and tritium will produce enhanced background rates for up to about 10 h (see Chap. 7). In Chap. 8 a very promising method to eliminate the problem arising from stored electrons in the main spectrometer was discussed which is based on stochastic heating of the stored electrons by the well known technique of Electron Cyclotron Resonance (ECR). This solution was successfully tested experimentally at the pre-spectrometer and its efficiency was extrapolated to the case of the main spectrometer by simulations.

Setting aside the very promising results of these investigations, other solutions to remove stored electrons are being pursued nevertheless. This is important, since it has not yet been demonstrated experimentally that the ECR method is efficient at the main spectrometer. Furthermore, scenarios involving possible technical complications related to the integrity of the wire electrode motivate the development of alternative methods. In the following three other techniques will be discussed in some detail.

C.1 Electric Dipole

The inner electrode systems of both pre- and main spectrometer are composed of two insulated half shells allowing to apply a static electric dipole field of up to $E = 100$ V/m. The electric field causes an $\vec{E} \times \vec{B}$-drift perpendicular to the dipole field (see Fig. C.1) which leads to a deformation of the electron's trajectory. However, the electron describes a closed path after a full magnetron cycle. If the electric dipole is strong enough, the deformation of the path is sufficient to guide the electron to an electrode surface where it is absorbed (see Fig. C.2). Simulations [5] and [6], however, showed that even with a dipole field of 100 V/m only electrons up to 2 keV can be removed.

S. Mertens, *Background Processes in the Electrostatic Spectrometers of the KATRIN Experiment*, Springer Theses, DOI: 10.1007/978-3-319-01177-6,

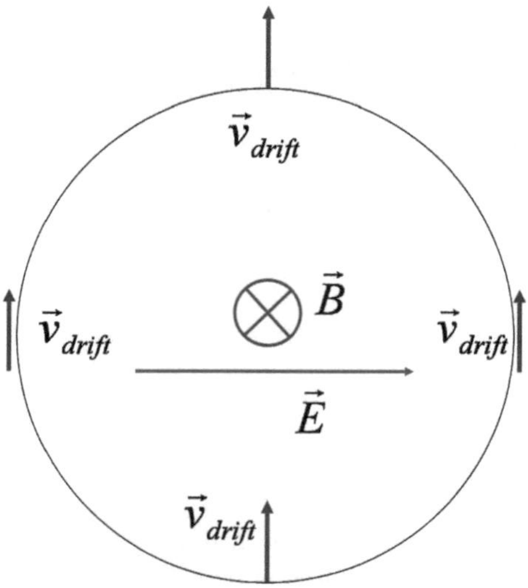

Fig. C.1 Drift velocity caused by a static electric dipole field

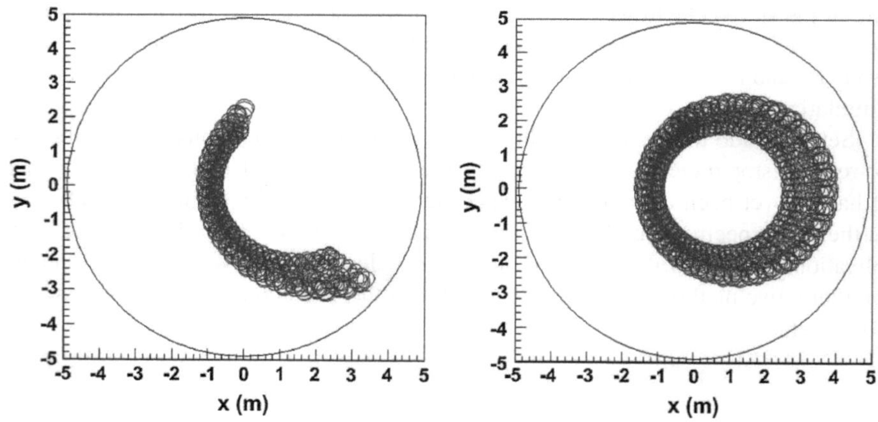

Fig. C.2 Trajectory of trapped electron in presence of an electric dipole field. *Left* The distortion of the trajectory is sufficient to eventually remove the electron. *Right* The electron remains trapped even in the presence of an electric dipole field [6]

C.2 Magnetic Pulse

Another method to remove stored electrons is based on the application of a magnetic pulse. Here, the magnetic pulse is to be understood as a fast reduction of the magnetic field inside the main spectrometer. This could be achieved with the LFCS coils. There are two basic ideas behind the effect of a magnetic pulse:

- A time dependent axial magnetic field induces an electric field according to the Maxwell Equations, $\nabla \times \vec{E} = -\frac{\partial B}{\partial t} \rightarrow \vec{E} = -\frac{r}{2}\frac{\partial B}{\partial t}$. This electric field subsequently causes a radial drift motion of the stored electrons, as shown in Fig. C.3.
- The magnetic field inside of the spectrometer is reduced to zero, so that the magnetic field lines of the flux tube touch the walls of the vessel. Consequently, all stored electrons are guided to the electrode surfaces and are removed.

The first effect is in principle the preferable, since one would rather avoid the field lines of the flux tube from touching the wall since when the field is increased again, particles located on the electrode surfaces are carried into the flux tube and could be stored again. However, this scenario needs to be tested experimentally.

To implement the removal method due to the first effect, however, a very rapidly decreasing magnetic field is necessary. Due to Lenz's law an eddy current is induced in the main spectrometer when the magnetic field outside of the spectrometer is changed, thus counteracting the change of the magnetic field. Calculations [7] show, the time to reduce the magnetic field from 3.5 to 2.5 G takes about 100 ms. Monte Carlo simulations, taking this constraint into account, [6] could show that feasible values of $\frac{\partial B}{\partial t}$ are too small to achieve large enough drift velocities to efficiently remove stored electrons.

As mentioned, using the magnetic pulse based on the second effect, bears the risk of increasing the background when increasing the magnetic field again to its nominal value. To circumvent this disadvantage a third option seems possible: The electrons collected from the electrode surface will be rather low in energy $E < 2$ keV. Therefore, it seems possible to combine the magnetic pulse with the static dipole. After the high-energy stored electrons would be removed with the magnetic pulse, the low-energy electrons, added after the pulse, could be removed by the static electric dipole.

An advantage of the magnetic pulse, possibly in combination with the electric dipole, is potentially the easier technical implementation. All component needed to apply both methods are already at hand at KATRIN. Nevertheless, the implications of continuously repeated magnetic pulses over a long-term measurement of many years have to be investigated in detail.

C.3 Static Pin

A third method has been proposed based on the installation of a large static pin in the main spectrometer which absorbs the trapped electrons. Figure C.4 shows how this pin could possibly be installed. Simulations [8, 9] show very effective removal properties of the pin. With the design shown in Fig. C.4 a reduction of stored electrons of up to about 90% could be reached. To improve its performance, the pin could be extended to a kind of "fishing rod", by attaching a string at the end of the pin. In this case the reduction can be increased to almost 100 %.

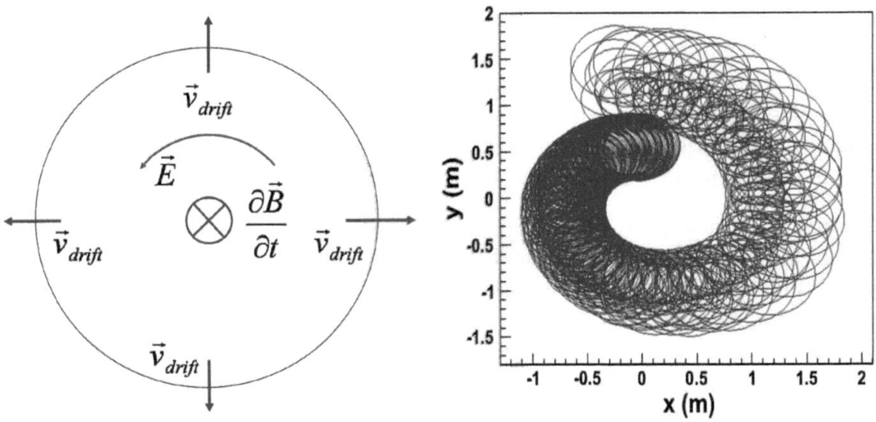

Fig. C.3 Working principle of magnetic pulse. *Left* The magnetic pulse induces a ring shaped electric field that leads to a radial drift motion of the stored electrons. Depending on the sign of $\frac{\partial B}{\partial t}$ the drift velocity is pointing inwards or outwards. *Right* The figure shows the trajectory of a stored electron in the presence of a magnetic pulse from its initial position close to the center of the flux tube [6]

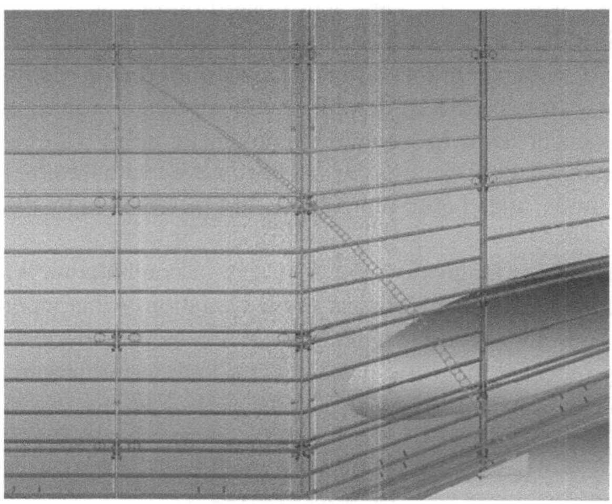

Fig. C.4 Technical drawing of a static pin in the main spectrometer

An obvious drawback of this method is that a part of the detector is shadowed by the pin, which reduces the count rate. Furthermore, the potential of the pin might influence the analyzing potential, however corresponding calculations reveal that the influence does not exceed acceptance limits. Finally, small angle scattering of electrons at the surface of the pin need to be understood in great detail in order to avoid systematic effects due to unaccounted energy losses.

Appendix D
Tables for Radon and Krypton Event Generator

In the following key physics parameters used by the radon generator (see Sect. 4.3.1) are given. Tables D.1, D.2 and D.3 show the probabilities for conversion electron creation for ^{219}Rn, ^{220}Rn and ^{83m}Kr decay, respectively. Table D.4 contains the probability for shake-off electron creation in radon α-decays. The probabilities for Auger electron emission following shake-off and conversions electron emission are taken from [10].

S. Mertens, *Background Processes in the Electrostatic Spectrometers of the KATRIN Experiment*, Springer Theses, DOI: 10.1007/978-3-319-01177-6,

Table D.1 Energy of emitted conversion electrons following α-decay of ^{219}Rn ([11]). The energy of the conversion electrons is given by the transition energy (last column) minus the binding energy of the shell (3rd column) that the conversion electron is emitted from. Only electron lines with an intensity larger than 0.05% are listed. The binding energies for the different shells in radon are as follows: K-shell - 93.1 keV, LI-shell - 16.939 keV, MI-shell - 4.1 keV

Electron energy [keV]	Probability [%]	Shell	Transition energy [keV]
37.5	0.4	K	130.6
113.7	0.13	L	130.6
178.13	1.27	K	271.2
254.29	0.74	L	271.2
267.08	0.19	M	271.2
270.24	0.06	NP	271.2
308.71	0.23	K	402
384.87	0.10	L	402

Table D.2 Energy of emitted conversion electrons following α-decay of ^{220}Rn ([12]). The probability for conversion electron emission in case of ^{220}Rn is negligibly small

Electron energy [keV]	Probability [%]	Shell	Transition energy [keV]
450	$3 \cdot 10^{-3}$	K	549.7
534	$6 \cdot 10^{-4}$	L	549.7

Table D.3 Energy of emitted conversion electrons following ^{83m}Kr decay ([13])

Electron energy [keV]	Probability [%]	Shell	Transition energy [keV]
7.484	0.80	L	32.15
9.117	0.13	M	32.15
9.381	0.01259	N	32.15
17.824	0.248	K	9.409
30.228	0.637	L	9.409
31.861	0.1069	M	9.409
32.125	0.00783	N	9.409

Table D.4 Shake off electron probabilities ([14, 15]). These data are based on measurements of ^{210}Po α-decay, which is the only isotope in the literature where shake-off electron data are available. However, the atom shells ^{210}Po, ^{220}Rn and ^{219}Rn are rather similar which justifies the use ^{210}Po here

Shell	Emission probability (%)
K	$2.2 \cdot 10^{-6}$
L	$5.0 \cdot 10^{-4}$
L2	$0.7 \cdot 10^{-4}$
L3	$1.6 \cdot 10^{-4}$
M 1,2,3,4	$4.0 \cdot 10^{-3}$

Appendix E
Simulation Input Parameters

E.1 Kassiopeia Configurations for Various Simulations

In chap. 7 several Monte Carlo simulation with KASSIOPEIA are presented. The detailed Kassiopeia configuration is given in the Tables E.1, E.2, E.3, E.4.

E.2 Electromagnetic Field Input Parameters

Here the typical electromagnetic input parameters to simulate the KATRIN beam line are given. The full input files can be found in the Kassiopeia subversion repository.

For most simulations of the main spectrometer in this work the two minima magnetic field configuration was chosen. Table E.5 shows the input parameters, where the WGTS, the DPS and CPS magnets and the earth magnetic field are approximated by large coils. For a subset of the simulations performed in this work the full model describing all 62 KATRIN magnets, including those partly tilted with respect to the beam axis, was used.

For simulations at the pre-spectrometer the magnetic field configuration given in Table E.6 was used where the amperturns were set to the value of interest.

The electric input parameter for the pre-spectrometer are given in table E.8. The actual input file comprising 294 input lines is much more sophisticated, the table thus shows a simplified configuration. In most simulations of this work the inner electrode was 500 V more negative than the tank. This setting forms a small Penning trap in the center of the pre-spectrometer. If this trap phenomenon was relevant for the analysis, simulations with both inner electrode on same potential and on more negative potential than the tank were performed.

The main spectrometer input parameters are given in Table E.7. As in the case of the pre-spectrometer only a reduced number of input parameters is listed.

S. Mertens, *Background Processes in the Electrostatic Spectrometers of the KATRIN Experiment*, Springer Theses, DOI: 10.1007/978-3-319-01177-6,

Table E.1 Kassiopeia simulation settings for comparison of simulated and measured rates at the pre-spectrometer (Sect. 7.2.3)

Simulation module	Parameter	Setting
KTRACK		
	Stepping mode	Exact
	Exit condition	$E_{\text{min}} = 50$ eV
		Hit electrode
		Leave spectrometer
	Synchrotron	On
	Scattering	$p_{H_2} = 4 \cdot 10^{-11}$ mbar
		$p_{H_2O} = 6 \cdot 10^{-11}$ mbar
KPAGE		
	Energy	^{219}Rn, ^{220}Rn $\alpha-$ decays
	Angular distribution	Isotropic
	Position	Homogeneous in sensitive pre-spectrometer volume
	Number of events	7,000 decays
Fields		
	Electric	Inner electrodes 500 V more negative (see table E.8)
	Magnetic	Full magnetic field (see Table E.6)

Table E.2 KASSIOPEIA simulation settings for comparison of storage probabilities in the pre- and main spectrometer (Sect. 7.3.1)

Simulation module	Parameter	Setting
KTRACK		
	Stepping mode	Exact
	Exit condition	Maximal time-of-flight
		Hit electrode
		Leave spectrometer
	Synchrotron	Off
	Scattering	Off
KPAGE		
	Energy	10 eV $< E_{\text{fix}} <$ 100 keV
	Angular distribution	Isotropic
	Position	Homogeneous in sensitive pre- and main spectrometer volume
	Number of events	1,100 electrons
Fields		
	Electric	PS: inner electrodes on tank potential (see Table E.8)
		MS: Table E.7
	Magnetic	PS: full magnetic field (see Table E.6)
		MS: One global minium (see Appendix E.2)

Table E.3 Kassiopeia simulation settings for energy dependence of number of secondaries (Sect. 7.3.1)

Simulation module	Parameter	Setting
KTRACK		
	Stepping mode	Exact
	Exit condition	$E_{\text{min}} = 13.6$ eV
		Hit electrode
		Leave spectrometer
	Synchrotron	Off
	Scattering	$p(H_2) = 10^{-11}$ mbar
KPAGE		
	Energy	$10\text{ eV} < E_{\text{fix}} < 100\text{ keV}$
	Angular distribution	Isotropic
	Position	Homogeneous in sensitive main spectrometer volume
	Number of events	9,000 electrons
Fields		
	Electric	Table E.7
	Magnetic	Two local magnetic field minima (see Table E.5)

Table E.4 Kassiopeia simulation settings background rate produced by a tritium β-decay and radon α-decay (Sect. 7.3.2)

Simulation module	Parameter	Setting
KTRACK		
	Stepping mode	Exact
	Exit condition	$E_{\text{min=13.6 eV}}$
		Hit electrode
		Leave spectrometer
	Synchrotron	Off
	Scattering	$p(H_2) = 10^{-11}$ mbar
KPAGE		
	Energy	^{219}Rn, ^{220}Rn $\alpha-$ decays, tritium β-decay
	Angular distribution	Isotropic
	Position	Homogeneous in sensitive main spectrometer volume
	Number of events	3,000 decays in total
Fields		
	Electric	Table E.7
	Magnetic	Two local magnetic field minima (see Table E.5)

Table E.5 Reference input parameters for the simulations based on a main spectrometer configuration with two magnetic field minima

Type	Position [m]	Amperturns [An]
Prespec magnet 1	−16.49	2120000
Prespec magnet 2	−12.13	2120000
Pinch magnet	12.22	3225000
Detector magnet	13.82	4230000
WGTS	−38.5	53100000
DPS 2	−27.2	29000000
CPS	−20	25170000
Earth magnetic field	0	−3246
Coil 1	−6.80	0
Coil 2	−4.95	50
Coil 3	−4.05	50
Coil 4	−3.15	50
Coil 5	−2.25	100
Coil 6	−1.35	220
Coil 7	−0.45	280
Coil 8	0.45	280
Coil 9	1.35	260
Coil 10	2.25	150
Coil 11	3.15	120
Coil 12	4.05	100
Coil 13	4.95	140
Coil 14 + 15	6.80	

Table E.6 Reference magnetic field input parameter for simulations of the pre-spectrometer

Type	Position [m]	Amperturns [An]
Prespec magnet 1	1.54	2120000
Prespec magnet 2	1.54	2120000

Table E.7 Reference electric field input parameter for simulations of the main spectrometer

Type	Potential [kV]
Tank	−18.4
Anti Penning electrode	−18.4
Outer cylindrical wire	−18.5
Inner cylindrical wire	−18.6
Outer conical wire	−18.5
Inner conical wire	−18.6
Outer steep cone wire	−18.56
Inner steep cone wire	−18.42

Table E.8 Reference electric field input parameter for simulations of the pre-spectrometer

Type	Potential [kV]
Tank	−18.0
West cone	−18.5 (18.0)
East cone	−18.5 (18.0)
Wire electrode	−18.5 (18.0)

Appendix F
Vacuum Setup of Pre- and Main Spectrometer

The pre-spectrometer vessel is an UHV recipient with a length of 3.4 m and a diameter of 1.7 m and a volume of $V_{PS} = 8.5$ m^3. Its vacuum system consists of cascaded turbo-molecular pumps (TMP) and a getter pump (SAES NEG strips St707). The first set of TMPs is two parallel Leybold MAG W 1,300 pumps connected to the pre-spectrometer vessel through tubes with a length of 100 cm and a diameter of 20 cm. These two tubes lead to a larger tube (pump port for the getter pump) with a length of 100 cm and a diameter of 50 cm, which ends in the pre-spectrometer vessel. The pre-spectrometer can be equipped with two getter pumps, one in each of the two pump ports. Each pump can be hold between 90 and 180m of NEG strips. The effective pumping speeds $S_{\text{eff}}^{\text{PS}}(\text{gas})$ of the pre-spectrometer vacuum system are listed in Table F.1.

Both spectrometers are connected by a beam tube with an in-line valve. The vacuum tube is 80 cm long with an inner diameter varying from 10 to 16 cm. The conductance for different molecules is given in Table F.1.

The main spectrometer is 23.3 m long with a diameter of 10 m corresponding to a volume of $V_{MS} = 1240$ m^3. Its vacuum system consists of 6 cascaded Leybold MAG W 2,800 TMPs connected to the main spectrometer vessel in parallel, and 3 getter pumps with a total pumping speed of 10^6 l/s for H_2. Since the getter material emanates radon, nitrogen cooled cryo-baffles have to be installed in front of the getter pumps, thus reducing the effective pumping speed to 45 % of the value given above. The effective pumping speeds of the main spectrometer vacuum system are listed in Table F.1.

S. Mertens, *Background Processes in the Electrostatic Spectrometers of the KATRIN Experiment*, Springer Theses, DOI: 10.1007/978-3-319-01177-6,

Table F.1 Parameters of the vacuum system. The table lists conductances (C) and pumping speeds (S) of different vacuum components of the pre- and main spectrometer for different gases at 20°C. The getter pumps have been simulated with a sticking coefficient of 2.8 [16, 17]. For the pumping speed of the cryo-baffles a sticking probability of $\alpha = 0.8$ was assumed

	N_2	Rn	H_2	HT	T_2
Molar mass	28	219	2	4	6
	pre-spectrometer				
$C_{20}(l = 100\text{ cm}, d = 20\text{ cm})$	765 l/s	274 l/s	2 862 l/s	2 024 l/s	1 653 l/s
$C_{50}(l = 100\text{ cm}, d = 50\text{ cm})$	9 075 l/s	3 245 l/s	33 956 l/s	24 010 l/s	19 604 l/s
Combined conductance C_{TMP}^{PS}	706 l/s	253 l/s	2 640 l/s	1 867 l/s	1 524 l/s
TMP pumping speed S_{TMP}^{PS}	1 100 l/s	640 l/s	920 l/s	1 210 l/s	1 280 l/s
S_{eff}^{PS}(2 TMPs)	860 l/s	360 l/s	1 360 l/s	1 470 l/s	1 390 l/s
S_{eff}^{PS}(90 m NEG)	4 000 l/s	0 l/s	28 500 l/s	20 200 l/s	16 500 l/s
S_{eff}^{PS}(180 m NEG)	5 000 l/s	0 l/s	33 400 l/s	23 600 l/s	19 300 l/s
S_{eff}^{PS}(1000 m NEG)	50 000 l/s	0 l/s	340 000 l/s	240 000 l/s	190 000 l/s
Cryo-baffle (50 cm)	0 l/s	26 500 l/s	0 l/s	0 l/s	0 l/s
	PS → MS beamline				
$C_{\text{PS}\rightarrow\text{MS}}$ (inline valve)	200 l/s	70 l/s	750 l/s	530 l/s	430 l/s
	Main spectrometer				
$C_{\text{TMP}}^{\text{MS}}$	2 650 l/s	950 l/s	9 930 l/s	7 020 l/s	5 730 l/s
TMP pumping speed $S_{\text{TMP}}^{\text{MS}}$	2 650 l/s	1 530 l/s	2 100 l/s	2 920 l/s	3 080 l/s
S_{eff}^{MS}(6 TMPs)	7 950 l/s	3 520 l/s	10 400 l/s	12 400 l/s	12 000 l/s
S_{eff}^{MS}(3000 m NEG, no baffle)	150 000 l/s	0 l/s	1 000 000 l/s	710 000 l/s	580 000 l/s
S_{eff}^{MS}(3000 m NEG, with baffle)	22 500 l/s	900 000 l/s	450 000 l/s	320 000 l/s	260 000 l/s

Appendix G
Details of Background Rate Arising from Stored Electrons

G.1 Analytic Derivation of Storage Probability

Within the adiabatic approximation one can calculate analytically the minimal transversal energy $E_{\perp}^{\text{min,trap}}$ for an electron to be trapped by a magnetic field $\vec{B}(\vec{p})$ at the position $\vec{p}$. The procedure can be illustrated by following consideration: An electron starting at the position $\vec{p}_{\text{start}}$ with kinetic energy $E_{\text{kin}}(\vec{p}_{\text{start}})$ has a kinetic energy of

$$E_{\text{kin}}(\vec{p}) = E_{\text{kin}}(\vec{p}_{\text{start}}) + \text{q}\,(U(\vec{p}_{\text{start}}) - U(\vec{p})) \tag{G.1}$$

at the position $\vec{p}$, where U denotes the electric potential and q the electric charge.

Suppose $E_{\text{kin}}(\vec{p})$ would be exclusively transversal energy $E_{\text{kin}}(\vec{p}) = E_{\perp}(\vec{p})$. In this case, the corresponding transversal energy at the starting position would be at the minimal value for which the electron would be stored:

$$E_{\perp}^{\text{min,trap}}(\vec{p}_{\text{start}}) = E_{\perp}(\vec{p}) \cdot \frac{|\vec{B}(\vec{p}_{\text{start}})|}{|\vec{B}(\vec{p})|} \tag{G.2}$$

If the condition $E_{\perp}^{\text{min,trap}}(\vec{p}_{\text{start}}) > E_{\text{kin}}(\vec{p}_{\text{start}})$ holds for all positions $\vec{p}$ the electron reaches, no storage is possible. Otherwise, the electron is stored. The question to answer then is: What is the minimal polar angle $\theta^{\text{min,trap}}$ of the electron at its starting point in order to procure the minimal transversal starting energy $E_{\perp}^{\text{min,trap}}(\vec{p}_{\text{start}})$? From a simple geometrical consideration follows

$$\theta^{\text{min,trap}} = \arcsin\left(\sqrt{\frac{E_{T}^{\text{min,trap}}(\vec{p}_{\text{start}})}{E_{\text{kin}}(\vec{p}_{\text{start}})}}\right). \tag{G.3}$$

Assuming an isotropic creation of the electrons, the probability P^{trap} for an electron to be produced with an angle larger than $\theta^{\text{min,trap}}$ is hence simply given by

S. Mertens, *Background Processes in the Electrostatic Spectrometers of the KATRIN Experiment*, Springer Theses, DOI: 10.1007/978-3-319-01177-6,

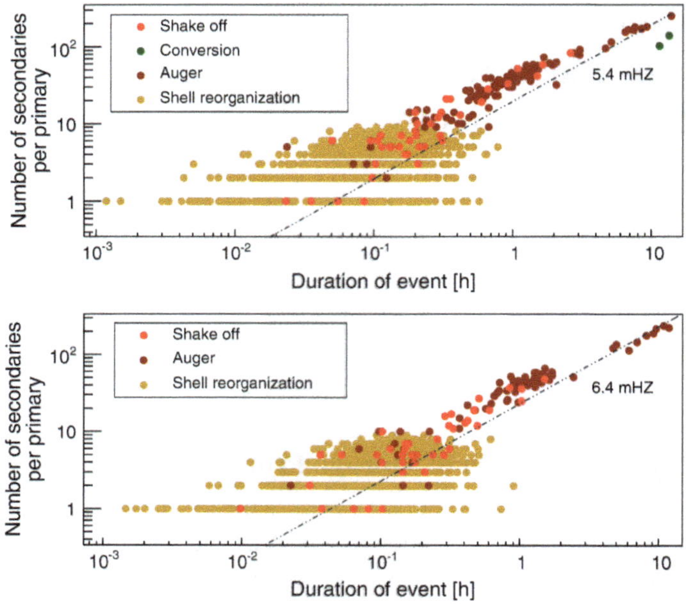

Fig. G.1 Rate produced by a ^{219}Rn and ^{220}Rn α-decays. This plot shows the number of secondary electrons produced by a single primary electron created in a radon α-decay versus its storage time. *Top* ^{219}Rn. *Bottom* ^{220}Rn

$$P^{\text{trap}} = \cos(\theta^{\text{min,trap}}) \tag{G.4}$$

G.2 Detailed Plots of Background Characteristics of ^{219}Rn and ^{220}Rn α-decays

In the following the background characteristics of ^{219}Rn and ^{220}Rn α-decays is shown in more detail. Figure G.1 shows the number of secondary electrons per individual primary electron together with the overall duration of the enhanced background level as a scatter plot. The plot allows to discriminate the different mechanisms leading to electron emission.

The second Fig. G.2 shows the number of secondaries and the duration of of the enhanced rate for a complete event. The color coding in this case indicates the multiplicity of the primary electrons emitted in the α-decay.

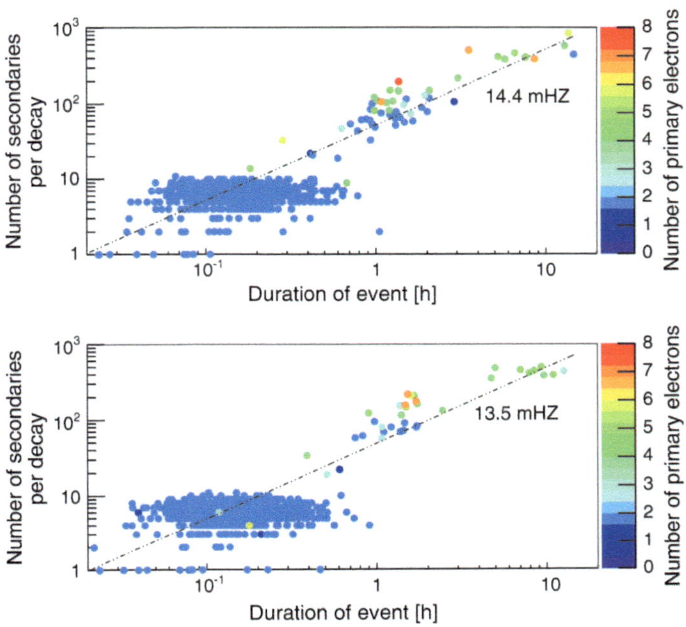

Fig. G.2 Dependence of rate produced by a ^{219}Rn and ^{220}Rn α-decays on number of primaries. This plot shows all events in which at least one secondary electron is produced. The color coding indicates how many primary electrons were produced in the decay. *Top* ^{219}Rn. *Bottom* ^{220}Rn

Curriculum Vitae

Personal Details

Susanne Mertens
2217 Stuart Street
Berkeley, CA 94705, USA
Telefon: +1 510 717 1876
E-mail: smertens@lbl.gov
Place of birth: Heidelberg
Date of birth: 25th May 1983

Current Position
Postdoc

Postdoc at the Institut for Nuclear and Astroparticle Physics at the Lawrence Berkeley National Laboratory (LBNL) Work at the Majorana und KATRIN Experiments Feodor Lynen Research Fellowship of the Alexander von Humboldt foundation	Aug. 2012—today Berkeley, USA

S. Mertens, *Background Processes in the Electrostatic Spectrometers of the KATRIN Experiment*, Springer Theses, DOI: 10.1007/978-3-319-01177-6,

Education
Ph.D.

Karlsruhe Tritium Neutrino (KATRIN) Experiment at the Institute of Experimental Nuclear Physics at the Karlsruhe Institute of Technology (KIT) — Aug. 2008–Feb. 2012, Karlsruhe, Germany

Thesis title: Investigation of Background Processes in the Electrostatic Spectrometers of the KATRIN Experiment

Erna-Scheffler thesis award of the Soroptimist International Club Karlsruhe

Participation in the European Center for Theoretical Physics (ECT*) Doctoral Training Program 2011 for Nuclear, Particle and Astrophysics — April 2011–June 2011, Trento, Italy

Research visit to Massachusetts Institute of Technology (MIT) — April 2010–July 2010, Cambridge, USA

Course of Studies

Study of Physics at the University of Karlsruhe — Oct. 2002–May 2008, Karlsruhe, Germany

Diploma thesis at the Institute for Theoretical Particle Physics Thesis title: The magnetic moment of the muon and the decay $\tau \rightarrow e\gamma$ in the MSSM at large tan β

Study abroad at Trinity College, Dublin — Oct. 2005–July 2006, Dublin, Ireland

School

High school: Dietrich Bonhoeffer Gymnasium — 1993–2002, Weinheim, Germany

Membership in bodies and Social Engagements

Elected member of the Scientific Technical Association (WTV) of the KIT Elementary Particle and Astroparticle Physics Center (KCETA) as young researcher	Feb. 2009—today
Representative of Ph.D students of the Graduate College High Energy Physics and Particle Astrophysics	April 2009–Oct. 2011
Active member of International Student Society (AK ERASMUS) at KIT	April 2007–April 2009

Susanne Mertens
Karlsruhe, March 23, 2013

References

1. N. Wandkowsky, Design and Background Simulations for the KATRIN Main Spectrometer and Air Coil System Master's thesis, KIT, 2009
2. M. Prall, Transmission properties of the pre-spectrometer at high surplus energies. PhD thesis, Universität Münster, 2011
3. P. Renschler, KESS: A new Monte Carlo simulation code for low-energy electron interactions in silicon detectors. PhD thesis, KIT, 2011
4. Z. Pei, C. Berglund, Angular distribution of Photoemission from Gold Thin Films, Jap. J. Appl. Phys. **41**(2), No 1A/B, (2002)
5. T. Thümmler,Entwicklung von Methoden zur Untergrundreduzierung am Mainzer Tritium beta Spektrometer, Johannes Gutenberg Universität Mainz, Master's thesis, 2002
6. N. Wandkowsky, Removing stored electrons by electric dipole and magnetic pulse, KATRIN Collaboration meeting March, 2011
7. E. Otten, Removing stored electrons by a magnetic pulse. Internal KATRIN report, 2010
8. S. Schnetzer, Optimierung der Feld- und Potentialberechnungen für das Drahtelektrodensystem des KATRIN-Hauptspektrometers, Master's thesis, KIT, 2011
9. B. Jung, Simulation des Myon-induzierten Untergrundes und dessen Speicherwahrscheinlichkeit am KATRIN Hauptspektrometer, Master's thesis, KIT, 2011
10. J. Baró, J. Sempau, J. Fernández-Varea, F. Salvat, PENELOPE: An algorithm for Monte Carlo simulation of the penetration and energy loss of electrons and positrons in matter. Nucl. Instrum. Meth. Phys. Res. Sect. B: Beam Interact. Mater. Atoms **100**(1), 31–46 (1995)
11. E. Browne, Nuclear Data Sheets for A = 215, 219, 223, 227, 231. Nucl. Data Sheets **93**(4), 763–1061 (2001)
12. S.-C. Wu, Nuclear Data Sheets for A = 216", Nuclear Data Sheets, vol. 108, no. 5, pp. 1057–1092, 2007.
13. S.-C. WU, Nuclear Data Sheets for A = 83, Nuclear Data Sheets, **92**,no. 4, pp. 893Ã¢â?¬â??1028, 2001.
14. M.S. Rapaport, F. Asaro, I. Perlman, K-shell electron shake-off accompanying alpha decay. Phys. Rev. C **11**, 1740–1745 (May 1975)
15. M.S. Rapaport, F. Asaro, I. Perlman, M- and L-shell electron shake-off accompanying alpha decay. Phys. Rev. C **11**, 1746–1754 (May 1975)

16. C. Day, X. Luo, A. Conte, A. Bonucci, P. Manini, Determination of the sticking probability of a Zr-V-Fe non-evaporable getter strip. J. Vacuum Sci. Technol A: Vacuum, Surf., Films **25**(4), 824–830 (2007)
17. X. Luo, L. Bornschein, C. Day, J. Wolf, KATRIN NEG pumping concept investigation, Vacuum **81**(6), 777–781, 2007. *Proceedings of the European Vacuum Conference (EVC-9)*

FSC
www.fsc.org
MIX
Papier aus verantwortungsvollen Quellen
Paper from responsible sources
FSC® C105338

If you have any concerns about our products,
you can contact us on
ProductSafety@springernature.com

In case Publisher is established outside the EU,
the EU authorized representative is:
Springer Nature Customer Service Center GmbH
Europaplatz 3, 69115 Heidelberg, Germany

Printed by Libri Plureos GmbH
in Hamburg, Germany